Encyclopaedia of
Nanotechnology

Encyclopaedia of
Nanotechnology

Vol. 8 : Electronics and Photovoltaics

Dr. Rita Khare

AGROTECH PRESS

JAIPUR ■ NEW DELHI

AGROTECH PRESS

Regd. Off. "Bony Residency", Gate No. 2
Opp. Tilak Public School, Vishweswariya Nagar,
Gopalpura Road, Jaipur - 302018, India
Ph: 0141 - 2761280, Telefax: 2761381
E-mail: agrotechpress@gmail.com

Delhi Office: 102, 1st Floor, Satyam House,
4327/3, Ansari Road, Darya Ganj,
New Delhi - 110002, India
Ph: 011-45652440
visit us: www.agrotechpress.com

Distributed by :
KEYA INTERNATIONAL PRESS LLP
E-mail: keyainternationalpressllp@gmail.com

Published in India, 2024

ISBN : 9789384568566 (10 Vol. Set)

Typeset by Agrotech Press

Printed and bound in India by Replika Press Pvt. Ltd.

Contents

Preface

With the constant drive to miniaturise electronics, much of modern technology – from computer chips to sensor devices – involves components on a minute scale. To manufacture components hundreds of times thinner than a human hair, companies require electrical engineers with a new set of skills. Functional nanotechnology, nanoelectronics, nanofabrication, manufacturing, quantum engineering and producing nanomaterials such as graphene and carbon nanotubes are all key skills covered in this domain. In common usage, nanotechnology refers to structures and materials and processes to fabricate them that exhibit useful properties resulting from sub-100nm features. Development of electronics with nanotechnology offers engineers a firm grounding in conventional electronics, plus the specialist skills at the electronics/ physics interface required to work at the forefront of modern nanoscale device fabrication. With the increasing demand for smaller, faster, and more highly integrated optical and electronic devices the fields of nanoelectronics and nanophotonics is emerging.

Nanoelectronics refer to the use of nanotechnology in electronic components. The term covers a diverse set of devices and materials, with the common characteristic that they are so small that inter-atomic interactions and quantum mechanical properties need to be studied extensively. Some of these candidates include: hybrid molecular/semiconductor electronics, one-dimensional nanotubes/nanowires, or advanced molecular electronics. Recent silicon CMOS technology generations, such as the 22nm node, are already within this regime. Nanoelectronics holds some answers for how we might increase the capabilities of electronics devices while we reduce their weight and power consumption. Some of the nanoelectronics areas under development, the book explore in more detail. Improving display screens on electronics devices is one, which involves reducing power consumption while decreasing the weight and thickness of the screens. Increasing the density of memory chips is another. Researchers are developing a type of memory chip with a projected density of one terabyte of memory per square inch or greater. Reducing the size of transistors used in integrated circuits is yet another application. Researchers believe it may be possible to put the power of all of today's present computers in the palm of one's hand.

Silicon has long been established as the material of choice for the microelectronics industry. This is not yet true in photonics, where the limited degrees of freedom in material design combined with the indirect bandgap are a major constraint. Recent developments, especially those enabled by nanoscale engineering of the electronic and photonic properties, are starting to change the picture, and some silicon nanostructures now approach or even exceed the performance of equivalent direct-bandgap materials. Focusing on the application area, namely photovoltaics, this volume review recent progress in silicon nanocrystals, nanowires and photonic crystals as key examples of functional nanostructures. The book assess the state of the art in each field and highlight the challenges that need to be overcome to make silicon a truly high-performing photonic material. The significant research interest in the engineering of photovoltaic structures at the nanoscale is directed toward enabling reductions in photovoltaic module fabrication and installation costs as well as improving cell power conversion efficiency. Nanostructured photovoltaic devices are defined as solar cells that consist of thin films (< 2 μm) of semiconductor material, incorporate nanoscale optoelectronic materials that have excitonic or quantum-confinement character and 4) can be fabricated entirely with low-temperature (< 300°C) processing steps. Especially the last point, the compatibility of many nanostructured PV materials with the use of low- temperature processing can enable the use of flexible, lightweight substrates, potentially leading to a lower cost of solar module fabrication and installation. With the emergence of a multitude of nanostructured photovoltaic device architectures, the question has arisen of where both the practical and the fundamental limits of performance reside in these new systems. A particular advantage of nanotechnology materials is the tunability of their optical and electronic properties, which enables improved photovoltaic power conversion efficiencies by implementing strategies for reducing thermal losses.

The volume focuses on our present understanding of the electrical conduction in single-molecule circuits and provides a thorough introduction to the experimental techniques and theoretical concepts of nanotechnology for developing electronic and photonic systems. In this sense, this publication will prove invaluable to both researchers and students interested in the field of nanoelectronics and nanoscience in general. By bridging the materials physics and chemistry at the atomic scale with device and system design, integration, and performance requirements, the chapters from this volume provide a coherent picture of theoretical and experimental research efforts and technology development in these highly interdisciplinary areas. The book can help to foster further research and cross-disciplinary interactions required to surmount the barriers facing future generations of technology design.

Dr. Rita Khare

1

Introduction

Nanoelectronics holds some answers for how we might increase the capabilities of electronics devices while we reduce their weight and power consumption. Some of the nanoelectronics areas under development, which you can explore in more detail by following the links provided in the next section, include the following topics. Improving display screens on electronics devices. This involves reducing power consumption while decreasing the weight and thickness of the screens.

Increasing the density of memory chips. Researchers are developing a type of memory chip with a projected density of one terabyte of memory per square inch or greater. Reducing the size of transistors used in integrated circuits. One researcher believes it may be possible to "put the power of all of today's present computers in the palm of your hand".

Nanoelectronics: Applications under Development

Researchers are looking into the following nanoelectronics projects:

- Cadmium selenide nanocrystals deposited on plastic sheets have been shown to form flexible electronic circuits. Researchers are aiming for a combination of flexibility, a simple fabrication process and low power requirements.
- Integrating silicon nanophotonics components into CMOS integrated circuits. This optical technique is intended to provide higher speed data transmission between integrated circuits than is possible with electrical signals.

- Researchers at UC Berkeley have demonstrated a low power method to use nanomagnets as switches, like transistors, in electrical circuits. Their method might lead to electrical circuits with much lower power consumption than transistor based circuits.
- Researchers at Georgia Tech, the University of Tokyo and Microsoft Research have developed a method to print prototype circuit boards using standard inkjet printers. Silver nanoparticle ink was used to form the conductive lines needed in circuit boards.
- Researchers at Caltech have demonstrated a laser that uses a nanopatterned silicon surface that helps produce the light with much tighter frequency control than previously achieved. This may allow much higher data rates for information transmission over fibre optics.
- Building transistors from carbon nanotubes to enable minimum transistor dimensions of a few nanometres and developing techniques to manufacture integrated circuits built with nanotube transistors.
- Researchers at Stanford University have demonstrated a method to make functioning integrated circuits using carbon nanotubes. In order to make the circuit work they developed methods to remove metallic nanotubes, leaving only semiconducting nanotubes, as well as an algorithm to deal with misaligned nanotubes. The demonstration circuit they fabricated in the university labs contains 178 functioning transistors.
- Developing a lead free solder reliable enough for space missions and other high stress environments using copper nanoparticles.
- Using electrodes made from nanowires that would enable flat panel displays to be flexible as well as thinner than current flat panel displays.
- Using semiconductor nanowires to build transistors and integrated circuits.
- Transistors built in single atom thick graphene film to enable very high speed transistors.
- Researchers have developed an interesting method of forming PN junctions, a key component of transistors, in graphene. They patterned the p and n regions in the substrate. When the graphene film was applied to the substrate electrons were either added or taken from the graphene, depending upon the doping of the substrate. The researchers believe that this method reduces the disruption of the graphene lattice that can occur with other methods.
- Combining gold nanoparticles with organic molecules to create a transistor known as a NOMFET (Nanoparticle Organic Memory Field-Effect Transistor).

- Using carbon nanotubes to direct electrons to illuminate pixels, resulting in a lightweight, millimetre thick "nanoemmissive" display panel.
- Using quantum dots to replace the fluorescent dots used in current displays. Displays using quantum dots should be simpler to make than current displays as well as use less power.
- Making integrated circuits with features that can be measured in nanometres (nm), such as the process that allows the production of integrated circuits with 22 nm wide transistor gates.
- Using nanosized magnetic rings to make Magnetoresistive Random Access Memory (MRAM) which research has indicated may allow memory density of 400 GB per square inch.
    - o Researchers have developed lower power, higher density method using nanoscale magnets called magnetoelectric random access memory (MeRAM).
- Developing molecular-sized transistors which may allow us to shrink the width of transistor gates to approximately one nm which will significantly increase transistor density in integrated circuits.
- Using self-aligning nanostructures to manufacture nanoscale integrated circuits.
- Using nanowires to build transistors without p-n junctions.
- Using buckyballs to build dense, low power memory devices.
- Using magnetic quantum dots in spintronic semiconductor devices. Spintronic devices are expected to be significantly higher density and lower power consumption because they measure the spin of electronics to determine a 1 or 0, rather than measuring groups of electronics as done in current semiconductor devices.
- Using nanowires made of an alloy of iron and nickel to create dense memory devices. By applying a current magnetized sections along the length of the wire. As the magnetized sections move along the wire, the data is read by a stationary sensor. This method is called race track memory.
- Using silver nanowires embedded in a polymer to make conductive layers that can flex, without damaging the conductor.
- IMEC and Nantero are developing a memory chip that uses carbon nanotubes. This memory is labelled NRAM for Nanotube-Based Nonvolatile Random Access Memory and is intended to be used in place of high density Flash memory chips.
- Researcher have developed an organic nanoglue that forms a nanometre thick film between a computer chip and a heat sink. They report that

using this nanoglue significantly increases the thermal conductance between the computer chip and the heat sink, which could help keep computer chips and other components cool.

- Researchers at Georgia Tech, the University of Tokyo and Microsoft Research have developed a method to print prototype circuit boards using standard inkjet printers. Silver nanoparticle ink was used to form the conductive lines needed in circuit boards.

Organic Electronics

Organic electronics is a field of material science concerning the design, synthesis, characterization, and application of organic small molecules or polymers that show desirable electronic properties such as conductivity. Unlike conventional inorganic conductors and semiconductors, organic electronic materials are constructed from organic (carbon-based) small molecules or polymers using synthetic strategies developed in the context of organic and polymer chemistry. One of the benefits of organic electronics is their low cost compared to traditional inorganic electronics.

Conductive materials are substances that can transmit electrical charges. Traditionally, most known conductive materials have been inorganic. Metals such as copper and aluminum are the most familiar conductive materials, and have high electrical conductivity due to their abundance of delocalized electrons that move freely throughout the inter-atomic spaces. Some metallic conductors are alloys of two or more metal elements, common examples of such alloys include steel, brass, bronze, and pewter.

In the eighteenth and early nineteenth centuries, people began to study the electrical conduction in metals. In his experiments with lightning, Benjamin Franklin proved that an electrical charge travels along a metallic rod. Later, Georg Simon Ohm discovered that the current passing through a substance is directly proportional to the potential difference, known as Ohm's law. This relationship between potential difference and current became a widely used measure of the ability of various materials to conduct electricity. Since the discovery of conductivity, studies have focused primarily on inorganic conductive materials with only a few exceptions.

Henry Letheby discovered the earliest known organic conductive material in 1862. Using anodic oxidation of aniline in sulfuric acid, he produced a partly conductive material, that was later identified as polyaniline. In the 1950s, the phenomenon that polycyclic aromatic compounds formed semi-conducting charge-transfer complex salts with halogens was discovered, showing that some organic compounds could be conductive as well.

More recent work has expanded the range of known organic conductive materials. A high conductivity of 1 S/cm (S = Siemens) was reported in 1963 for a derivative of tetraiodopyrrole. In 1972, researchers found metallic conductivity (conductivity comparable to a metal) in the charge-transfer complex TTF-TCNQ.

In 1977, it was discovered that polyacetylene can be oxidized with halogens to produce conducting materials from either insulating or semiconducting materials. In recent decades, research on conductive polymers has prospered, and the 2000 Nobel Prize in Chemistry was awarded to Alan J. Heeger, Alan G. MacDiarmid, and Hideki Shirakawa jointly for their work on conductive polymers. Conductive plastics have recently undergone development for applications in industry. In 1987, the first organic diode device of was produced at Eastman Kodak by Ching W. Tang and Steven Van Slyke. Spawning the field of organic light-emitting diodes (OLED) research and device production. For his work, Ching W. Tang is widely considered as the father of organic electronics for this work.

Technology for plastic electronics constructed on thin and flexible plastic substrates was developed in the 1990s. In 2000, the company Plastic Logic was founded as a spin-off of Cavendish Laboratory to develop a broad range of products using the plastic electronics technology.

Conductive Organic Materials

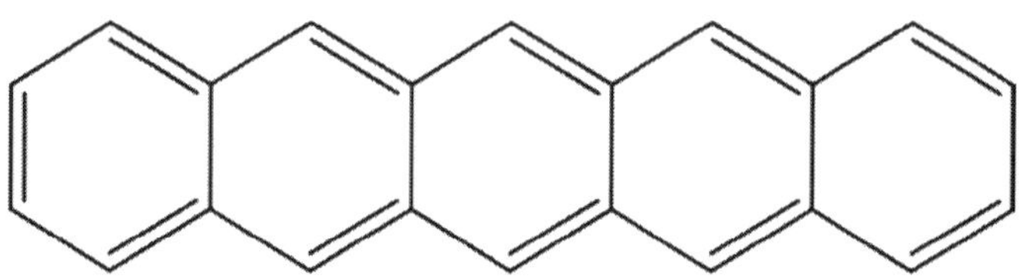

Pentacene

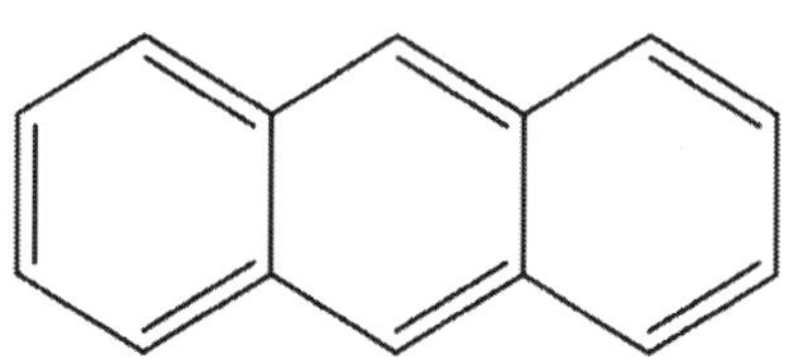

Anthracene

Figure: *Typical semiconducting small molecules*

Attractive properties of polymer conductors include a wide range of electrical conductivity that, can be tuned by varying the concentrations of chemical dopants, mechanical flexibility, and high thermal stability. Organic conductive materials can be grouped into two main classes: conductive polymers and conductive small molecules.

Conductive small molecules are usually used in the construction of organic semiconductors, which exhibit degrees of electrical conductivity between those of insulators and metals. Semiconducting small molecules include polycyclic aromatic compounds such as pentacene, anthracene and rubrene.

Conductive polymers are typically intrinsically conductive. Their conductivity can be comparable to metals or semiconductors. Most conductive polymers are not thermoformable, during production. However they can provide very high electrical conductivity without showing similar mechanical properties to other commercially available polymers. Both organic synthesis and advanced dispersion techniques can be used to tune the electrical properties of conductive polymers, unlike typical inorganic conductors. The most well-studied class of conductive polymers is the so-called linear-backbone "polymer blacks" including polyacetylene, polypyrrole, polyaniline, and their copolymers. Poly(p-phenylene vinylene) and its derivatives are used for electroluminescent semiconducting polymers. Poly(3-alkythiophenes) are also a typical material for use in solar cells and transistors.

Organic Light-Emitting Diode

An OLED (organic light-emitting diode) consists of a thin film of organic material that emits light under stimulation by an electric current. A typical OLED consists of an anode, a cathode, OLED organic material and a conductive layer.

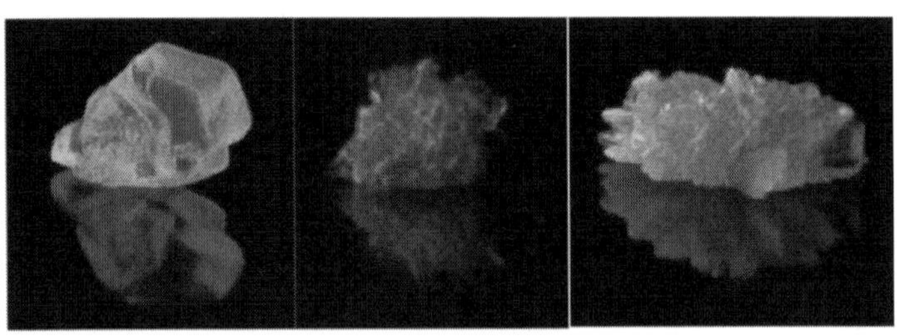

Figure: *Br6A, a next generation pure organic light emitting crystal family*

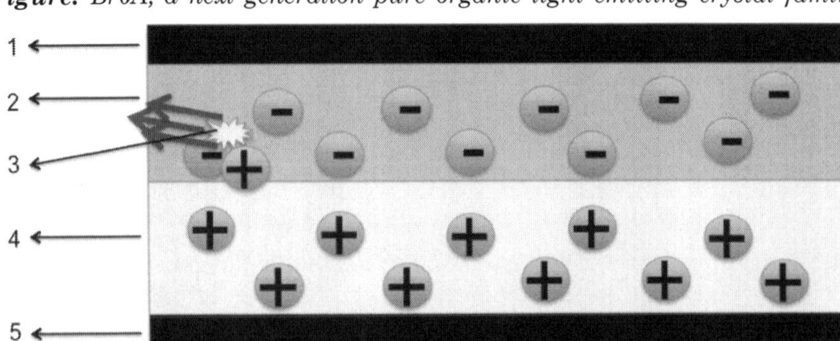

Figure: *Schematic of a bilayer OLED: 1. Cathode ("), 2. Emissive layer, 3. Emission of radiation, 4. Conductive layer, 5. Anode (+)*

Discovery of OLED

André Bernaose was the first to observe electroluminescence in organic materials, and Ching W. Tang, reported fabrication of an OLED device in 1987. The OLED device incorporated a double-layer structure motif consisting of separate hole transporting and electron-transporting layers, with light emission taking place in between the two layers. Their discovery opened a new era of current OLED research and device design.

Classification and Current Research Progress

OLED organic materials can be divided into two major families: small-molecule-based and polymer-based. Small molecule OLEDs (SM-OLEDs) include organometallic chelates(Alq3), fluorescent and phosphorescent dyes, and conjugated dendrimers. Fluorescent dyes can be selected according to the desired range of emission wavelengths; compounds like perylene and rubrene are often used. Very recently, Dr. Kim J. et al. at University of Michigan reported a pure organic light emitting crystal, Br6A, by modifying its halogen bonding, they succeeded in tuning the phosphorescence to different wavelengths including green, blue and red. By modifying the structure of Br6A, scientists are attempting to achieve a next generation organic light emitting diode. Devices based on small molecules are usually fabricated by thermal evaporation under vacuum. While this method enables the formation of well-controlled homogeneous film; is hampered by high cost and limited scalability.

Polymer light-emitting diodes (PLEDs), similar to SM-OLED, emit light under an applied electrical current. Polymer-based OLEDs are generally more efficient than SM-OLEDs requiring a comparatively lower amount of energy to produce the same luminescence. Common polymers used in PLEDs include derivatives of poly(p-phenylene vinylene) and polyfluorene. The emitted colour can be tuned by substitution of different side chains onto the polymer backbone or modifying the stability of the polymer. In contrast to SM-OLEDs, polymer-based OLEDs cannot be fabricated through vacuum evaporation, and must instead be processed using solution-based techniques. Compared to thermal evaporation, solution based methods are more suited to creating films with large dimensions. Zhenan Bao. et al. at Stanford University reported a novel way to construct large-area organic semiconductor thin films using aligned single crystalline domains.

Organic Field-Effect Transistor

An organic field-effect transistor (OFET) is a field-effect transistor using an organic semiconductor in its channel. OFETs can be prepared either by vacuum evaporation of small molecules, by solution-casting of polymers or small molecules, or by mechanical transfer of a peeled single-crystalline organic layer onto a substrate. These devices have been developed to realise low-cost, large-area electronic products and biodegradable electronics. OFETs have been fabricated with various device geometries. The most commonly used device geometry is bottom gate with top drain and source electrodes, because this geometry is similar to the thin-film silicon transistor (TFT) using thermally grown SiO_2 as gate dielectric. Organic polymers, such as poly(methyl-methacrylate) (PMMA), can also be used as dielectric.

In May 2007, Sony reported the first full-colour, video-rate, flexible, all plastic display, in which both the thin-film transistors and the light-emitting pixels were made of organic materials.

Materials

One common feature of OFET materials is the inclusion of an aromatic or otherwise conjugated π-electron system, facilitating the delocalization of orbital wavefunctions. Electron withdrawing groups or donating groups can be attached that facilitate hole or electron transport.

OFETs employing many aromatic and conjugated materials as the active semiconducting layer have been reported, including small molecules such as rubrene, tetracene, pentacene, diindenoperylene, perylenediimides, tetracyanoquinodimethane (TCNQ), and polymers such as polythiophenes (especially poly(3-hexylthiophene) (P3HT)), polyfluorene, polydiacetylene, poly(2,5-thienylene vinylene), poly(p-phenylene vinylene) (PPV).

The field is very active, with newly synthesized and tested compounds reported weekly in prominent research journals. Many review articles exist documenting the development of these materials. Rubrene-based OFETs show the highest carrier mobility 20–40 $cm^2/(V \cdot s)$. Another popular OFET material is pentacene, which has been used since the 1980s, but with mobilities 10 to 100 times lower mobilities (depending on the substrate) than rubrene. The major problem with pentacene, as well as many other organic conductors, is its rapid oxidation in air to form pentacene-quinone. However if the pentacene is preoxidized, and the thus formed pentacene-quinone is used as the gate insulator, then the mobility can approach the rubrene values. This pentacene oxidation technique is akin to the silicon oxidation used in the silicon electronics.

Polycrystalline tetrathiafulvalene and its analogues result in mobilities in the range 0.1–1.4 $cm^2/(V \cdot s)$. However, the mobility exceeds 10 $cm^2/(V \cdot s)$ in solution-grown or vapor-transport-grown single crystalline hexamethylene-tetrathiafulvalene (HMTTF). The ON/OFF voltage is different for devices grown by those two techniques, presumably due to the higher processing temperatures using in the vapor transport grows.

All the above-mentioned devices are based on p-type conductivity. N-type OFETs are yet poorly developed. They are usually based on perylenediimides or fullerenes or their derivatives, and show electron mobilities below 2 $cm^2/(V \cdot s)$.

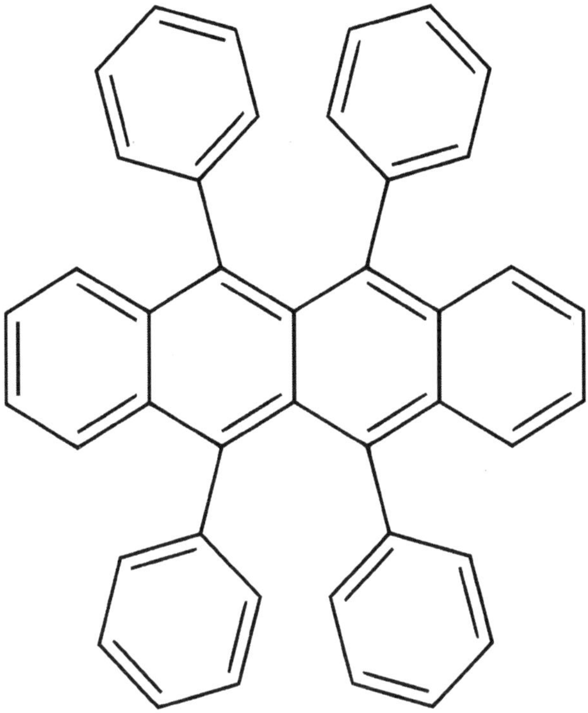

Figure: *Rubrene-OFET with the highest charge mobility*

An Organic field-effect transistor is a field-effect transistor utilising organic molecules or polymers as the active semiconducting layer. A field-effect transistor(FET) is any semiconductor material that utilises electric field to control the shape of a channel of one type of charge carrier, thereby changing its conductivity. Two major classes of FET are n-type and p-type semiconductor, classified according to the charge type carried. In the case of organic FETs (OFETs), p-type OFET compounds are generally more stable than n-type due to the susceptibility of the latter to oxidative damage.

Discovery of the OFET

J.E. Lilienfeld first proposed the field-effect transistor in 1930, but the first OFET was not reported until 1987, when Koezuka et al. constructed one using Polythiophene which shows extremely high conductivity. Other conductive polymers have been shown to act as semiconductors, and newly synthesized and characterized compounds are reported weekly in prominent research journals. Many review articles exist documenting the development of these materials.

Classification of OFET and Current Research Progress

Like OLEDs, OFETs can be classified into small-molecule and polymer-based system. Charge transport in OFETs can be quantified using a measure called carrier mobility; currently, rubrene-based OFETs show the highest carrier mobility of 20–40 $cm^2/(V \cdot s)$. Another popular OFET material is Pentacene. Due to its low solubility in most organic solvents, it's difficult to fabricate thin film transistors (TFTs) from pentacene itself using conventional spin-cast or, dip coating methods, but this obstacle can be overcome by using the derivative TIPS-pentacene. Current research focuses more on thin-film transistor (TFT) model, which eliminates the usage of conductive materials. Very recently, two studies conducted by Dr. Bao Z. et al. and Dr. Kim J. et al. demonstrated control over the formation of designed thin-film transistors. By controlling the formation of crystalline TFT, it is possible to create an aligned (as opposed to randomly ordered) charge transport pathway, resulting in enhanced charge mobility.

Organic Electronic Devices

Organic solar cells could cut the cost of solar power by making use of inexpensive organic polymers rather than the expensive crystalline silicon used in most solar cells. What's more, the polymers can be processed using low-cost equipment such as ink-jet printers or coating equipment employed to make photographic film, which reduces both capital and operating costs compared with conventional solar-cell manufacturing.

Silicon thin film solar cells on flexible substrates allow a significant cost reduction of large-area photovoltaics for several reasons:

1. The so-called 'roll-to-roll'-deposition on flexible sheets is much easier to realise in terms of technological effort than deposition on fragile and heavy glass sheets.
2. Transport and installation of lightweight flexible solar cells also saves cost as compared to cells on glass.

Inexpensive polymeric substrates like polyethylene terephthalate (PET) or polycarbonate (PC) have the potential for further cost reduction in photovoltaics. Protomorphous solar cells prove to be a promising concept for efficient and low-cost photovoltaics on cheap and flexible substrates for large-area production as well as small and mobile applications.

One advantage of printed electronics is that different electrical and electronic components can be printed on top of each other, saving space and increasing reliability and sometimes they are all transparent. One ink must not damage another, and low temperature annealing is vital if low-cost flexible materials such as paper and plastic film are to be used. There is much sophisticated engineering and chemistry involved here, with iTi, Pixdro, Asahi Kasei, Merck & Co. | Merck, BASF, HC Starck, Hitachi Chemical and Frontier Carbon Corporation among the leaders. Electronic devices based on organic compounds are now widely used, with many new products under development. Sony reported the first full-colour, video-rate, flexible, plastic display made purely of organic materials; television screen based on OLED materials; biodegradable electronics based on organic compound and low-cost organic solar cell are also available.

Fabrication Methods

There are important differences between the processing of small molecule organic semiconductors and semiconducting polymers. Small molecule semiconductors are quite often insoluble and typically require deposition via vacuum sublimation. While usually thin films of soluble conjugated polymers. Devices based on conductive polymers can be prepared by solution processing methods. Both solution processing and vacuum based methods produce amorphous and polycrystalline films with variable degree of disorder. "Wet" coating techniques require polymers to be dissolved in a volatile solvent, filtered and deposited onto a substrate. Common examples of solvent-based coating techniques include drop casting, spin-coating, doctor-blading, inkjet printing and screen printing. Spin-coating is a widely used technique for small area thin film production. It may result in a high degree of material loss. The doctor-blade technique results in a minimal material loss and was primarily developed for large area thin film production. Vacuum based thermal deposition of small molecules requires evaporation of molecules from a hot source. The molecules are then transported through vacuum onto a substrate. The process of condensing these molecules on the substrate surface results in thin film formation. Wet coating techniques can in some cases be applied to small molecules depending on their solubility.

Organic Solar Cell

Compared to conventional inorganic solar cell, organic solar cells have the advantage of lower fabrication cost. An organic solar cell is a device that uses organic electronics to convert light into electricity. Organic solar cells utilise organic photovoltaic materials, organic semiconductor diodes that convert light into electricity. Figure to the right shows five commonly used organic photovoltaic materials. Electrons in these organic molecules can be delocalized in a delocalized π orbital with a corresponding π^* antibonding orbital. The difference in energy between the π orbital, or highest occupied molecular orbital(HOMO), and π^* orbital, or lowest unoccupied molecular orbital(LUMO) is called the band gap of organic photovoltaic materials. Typically, the band gap lies in the range of 1-4eV.

The difference in the band gap of organic photovoltaic materials leads to different chemical structures and forms of organic solar cells. Different forms of solar cells includes single-layer organic photovoltaic cells, bilayer organic photovoltaic cells and heterojunction photovoltaic cells. However, all three of these types of solar cells share the approach of sandwiching the organic electronic layer between two metallic conductors, typically indium tin oxide.

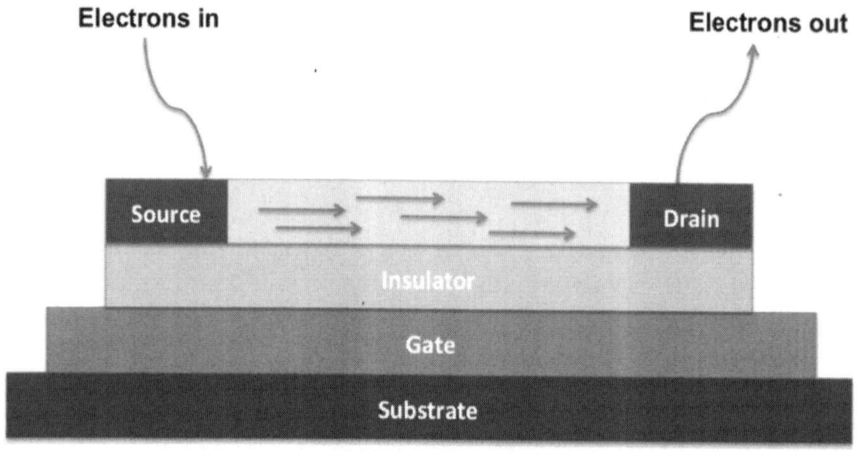

Figure: *Illustration of thin film transistor device*

Device of Organic Field-Effect Transistors

An organic field-effect transistor device consists of three major components: the source, the drain and the gate. Generally, a field-effect transistor has two plates, source in contact with drain and the gate respectively, working as conducting channel. The electrons move from source to the drain, and the gate serves to control the electrons' movement from source to drain. Different types of FETs are designed based on carrier properties. Thin film transistor (TFT), among them, is an easy fabricating one. In a thin film transistor, the source and drain are made by directly depositing a thin layer of semiconductor followed by a thin film of insulator between semiconductor and the metal gate contact. Such a thin film is made by either thermal evaporation, or simply spins coating. In a TFT device, there is no carrier movement between the source and drain. After applying a positive charge, accumulation of electrons on the interface cause bending of the semiconductor and ultimately lowers the conduction band with regards to the Fermi-level of the semiconductor. Finally, a highly conductive channel is formed at the interface.

Features

Conductive polymers are lighter, more flexible, and less expensive than inorganic conductors. This makes them a desirable alternative in many applications. It also creates the possibility of new applications that would be impossible using copper or silicon.

Organic electronics not only includes organic semiconductors, but also organic dielectrics, conductors and light emitters.

New applications include smart windows and electronic paper. Conductive polymers are expected to play an important role in the emerging science of molecular computers.

In general organic conductive polymers have a higher resistance and therefore conduct electricity poorly and inefficiently, as compared to inorganic conductors. Researchers currently are exploring ways of "doping" organic semiconductors, like melanin, with relatively small amounts of conductive metals to boost conductivity. Another method is to rely on two layers of gate dielectric, thereby allowing the high turn-on voltage caused by the fluorinated polymer to be mitigated by the low-voltage electrical characteristics of the high-k metal-oxide and the low defect count at the interface of the former to be used instead of the leaky defects of the latter. However, for many applications, inorganic conductors will remain the only viable option.

Flexible Organic SingleCrystal FieldEffect Transistors

An organic field-effect transistor (OFET) is a field-effect transistor using an organic semiconductor in its channel. OFETs can be prepared either by vacuum evaporation of small molecules, by solution-casting of polymers or small molecules, or by mechanical transfer of a peeled single-crystalline organic layer onto a substrate. These devices have been developed to realise low-cost, large-area electronic products and biodegradable electronics. OFETs have been fabricated with various device geometries. The most commonly used device geometry is bottom gate with top drain and source electrodes, because this geometry is similar to the thin-film silicon transistor (TFT) using thermally grown SiO_2 as gate dielectric. Organic polymers, such as poly(methyl-methacrylate) (PMMA), can also be used as dielectric.

In May 2007, Sony reported the first full-colour, video-rate, flexible, all plastic display, in which both the thin-film transistors and the light-emitting pixels were made of organic materials.

History of OFETs

The field-effect transistor (FET) was first proposed by J.E. Lilienfeld, who received a patent for his idea in 1930. He proposed that a field-effect transistor behaves as a capacitor with a conducting channel between a source and a drain electrode. Applied voltage on the gate electrode controls the amount of charge carriers flowing through the system.

The first field-effect transistor was designed and prepared in 1960 by Kahng and Atalla using a metal–oxide–semiconductor (MOSFET). However, rising costs of materials and manufacturing, as well as public interest in more environmentally friendly electronics materials have supported development of organic based electronics in more recent years. In 1987, Koezuka and co-workers reported the first organic field-effect transistor based on a polymer of thiophene molecules. The thiophene polymer is a type of conjugated polymer that is able to conduct charge, eliminating the need to use expensive metal oxide semiconductors. Additionally, other conjugated polymers have been shown to have semiconducting properties. OFET design has also improved in the past few decades. Many OFETs are now designed based on the thin-film transistor (TFT) model, which allows the devices to use less conductive materials in their design. Improvement on these models in the past few years have been made to field-effect mobility and on–off current ratios.

Materials

One common feature of OFET materials is the inclusion of an aromatic or otherwise conjugated π-electron system, facilitating the delocalization of orbital wavefunctions. Electron withdrawing groups or donating groups can be attached that facilitate hole or electron transport.

OFETs employing many aromatic and conjugated materials as the active semiconducting layer have been reported, including small molecules such as rubrene, tetracene, pentacene, diindenoperylene, perylenediimides, tetracyanoquinodimethane (TCNQ), and polymers such as polythiophenes (especially poly(3-hexylthiophene) (P3HT)), polyfluorene, polydiacetylene, poly(2,5-thienylene vinylene), poly(p-phenylene vinylene) (PPV).

The field is very active, with newly synthesized and tested compounds reported weekly in prominent research journals. Many review articles exist documenting the development of these materials.

Rubrene-based OFETs show the highest carrier mobility 20–40 $cm^2/(V \cdot s)$. Another popular OFET material is pentacene, which has been used since the 1980s, but with mobilities 10 to 100 times lower mobilities (depending on the substrate) than rubrene. The major problem with pentacene, as well as many other organic conductors, is its rapid oxidation in air to form pentacene-quinone. However if the pentacene is preoxidized, and the thus formed pentacene-quinone is used as the gate insulator, then the mobility can approach the rubrene values. This pentacene oxidation technique is akin to the silicon oxidation used in the silicon electronics.

Polycrystalline tetrathiafulvalene and its analogues result in mobilities in the range 0.1–1.4 $cm^2/(V \cdot s)$. However, the mobility exceeds 10 $cm^2/(V \cdot s)$ in solution-grown or vapor-transport-grown single crystalline hexamethylene-tetrathiafulvalene (HMTTF). The ON/OFF voltage is different for devices grown by those two techniques, presumably due to the higher processing temperatures using in the vapor transport grows. All the above-mentioned devices are based on p-type conductivity. N-type OFETs are yet poorly developed. They are usually based on perylenediimides or fullerenes or their derivatives, and show electron mobilities below 2 $cm^2/(V \cdot s)$.

Device Design of Organic Field-Effect Transistors

Three essential components of field-effect transistors are the source, the drain and the gate. Field-effect transistors usually operate as a capacitor. They are composed of two plates. One plate works as a conducting channel between two ohmic contacts, which are called the source and the drain contacts. The other plate works to control the charge induced into the channel, and it is called the gate. The direction of the movement of the carriers in the channel is from the source to the drain. Hence the relationship between these three components is that the gate controls the carrier movement from the source to the drain.

When this capacitor concept is applied to the device design, various devices can be built up based on the difference in the controller – i.e. the gate. This can be the gate material, the location of the gate with respect to the channel, how the gate is isolated from the channel, and what type of carrier is induced by the gate voltage into channel (such as electrons in an n-channel device, holes in a p-channel device, and both electrons and holes in a double injection device).

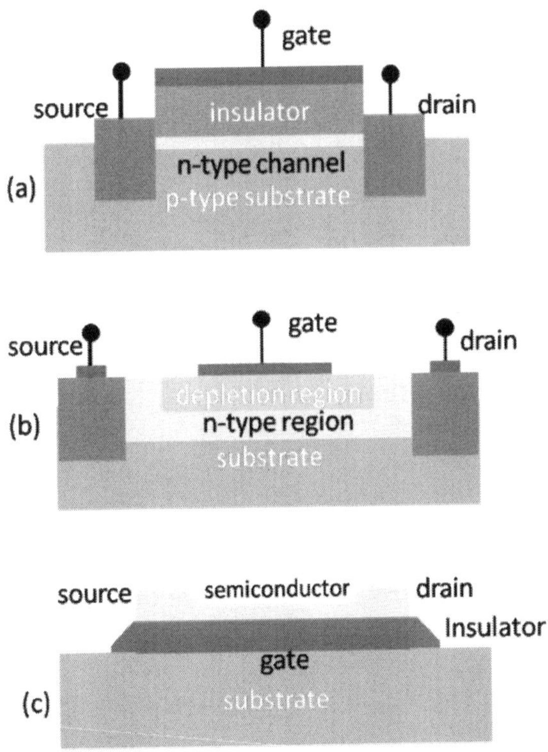

Figure: *Schematic of three kinds of field-effect transistor (FET): (a) metal-insulator-semiconductor FET (MISFET); (b) metal-semiconductor FET (MESFET); (c) thin-film transistor (TFT).*

Classified by the properties of the carrier, three types of FETs are shown schematically in Figure 1. They are MOSFET (metal–oxide–semiconductor field-effect transistor), MESFET (metal–semiconductor field-effect transistor) and TFT (thin-film transistor).

Misfet

The most prominent and widely used FET in modern microelectronics is the MOSFET. There are different kinds in this category, such as MISFET (metal–insulator–semiconductor field-effect transistor), and IGFET (insulated-gate FET). The source and the drain are connected by a semiconductor and the gate is separated from the channel by a layer of insulator. If there is no bias (potential difference) applied on the gate, the band bending is induced due to the energy difference of metal conducting band and the semiconductor Fermi level. Therefore a higher concentration of holes is formed on the interface of the semiconductor and the insulator.

When an enough positive bias is applied on the gate contact, the bended band becomes flat. If a larger positive bias is applied, the band bending in the opposite direction occurs and the region close to the insulator-semiconductor interface becomes depleted of holes. Then the depleted region is formed. At an even larger positive bias, the band bending becomes so large that the Fermi level at the interface of the semiconductor and the insulator becomes closer to the bottom of the conduction band than to the top of the valence band, therefore, it forms an inversion layer of electrons, providing the conducting channel. Finally, it turns the device on.

Mesfet

The only difference of this one from the MISFET is that the n-type source and drain are connected by an n-type region. In this case, the depletion region extends all over the n-type channel at zero gate voltage in a normally "off" device (it is similar to the larger positive bias in MISFET case). In the normally "on" device, a portion of the channel is not depleted, and thus leads to passage of a current at zero gate voltage.

TFT

The concept of TFT was first proposed by Paul Weimer in 1962. Here the source and drain electrodes are directly deposited onto the conducting channel (a thin layer of semiconductor) then a thin film of insulator is deposited between the semiconductor and the metal gate contact. This structure suggests that there is no depletion region to separate the device from the substrate. If there is zero bias, the electrons are expelled from the surface due to the Fermi-level energy difference of the semiconductor and the metal. This leads to band bending of semiconductor. In this case, there is no carrier movement between the source and drain. When the positive charge is applied, the accumulation of electrons on the interface leads to the bending of the semiconductor in an opposite way and leads to the lowering of the conduction band with regards to the Fermi-level of the semiconductor. Then a highly conductive channel forms at the interface.

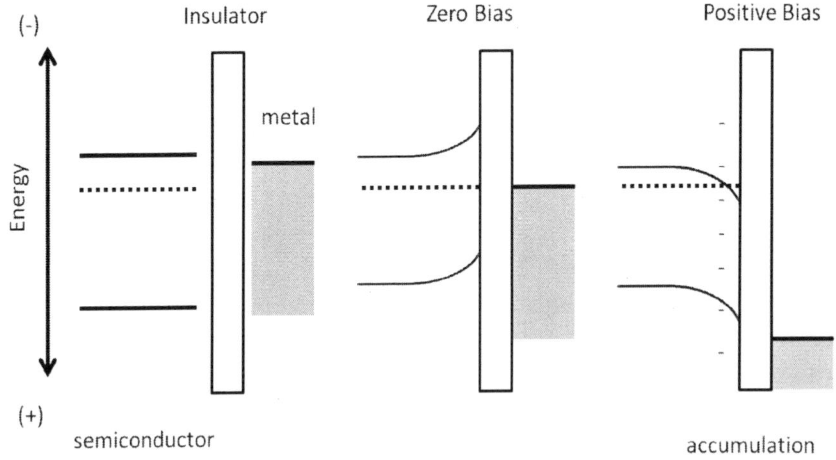

Figure: *Schematic of band-bending in the TFT device model.*

OFET

OFETs adopt the architecture of TFT. With the development of the conducting polymer, the semiconducting properties of small conjugated molecules have been recognised. The interest in OFETs has grown enormously in the past ten years. The reasons for this surge of interest are manifold. The performance of OFETs, which can compete with that of amorphous silicon (a-Si) TFTs with field-effect mobilities of 0.5–1 $cm^2 V^{-1} s^{-1}$ and ON/OFF current ratios (which indicate the ability of the device to shut down) of 10^6–10^8, has improved significantly. Currently, thin-film OFET mobility values of 5 $cm^2 V^{-1} s^{-1}$ in the case of vacuum-deposited small molecules and 0.6 $cm^2 V^{-1} s^{-1}$ for solution-processed polymers have been reported. As a result, there is now a greater industrial interest in using OFETs for applications that are currently incompatible with the use of a-Si or other inorganic transistor technologies. One of their main technological attractions is that all the layers of an OFET can be deposited and patterned at room temperature by a combination of low-cost solution-processing and direct-write printing, which makes them ideally suited for realisation of low-cost, large-area electronic functions on flexible substrates.

Device Preparation

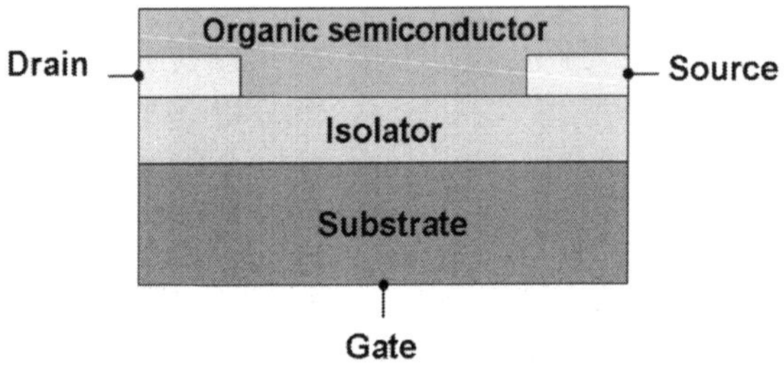

Figure: *OFET schematic*

Thermally oxidized silicon is a traditional substrate for OFETs where the silicon dioxide serves as the gate insulator. The active FET layer is usually deposited onto this substrate using either (i) thermal evaporation, (ii) coating from organic solution, or (iii) electrostatic lamination. The first two techniques result in polycrystalline active layers; they are much easier to produce, but result in relatively poor transistor performance. Numerous variations of the solution coating technique (ii) are known, including dip-coating, spin-coating, inkjet printing and screen printing.

The electrostatic lamination technique is based on manual peeling of a thin layer off a single organic crystal; it results in a superior single-crystalline active layer, yet it is more tedious. The thickness of the gate oxide and the active layer is below one micrometre.

Carrier Transport

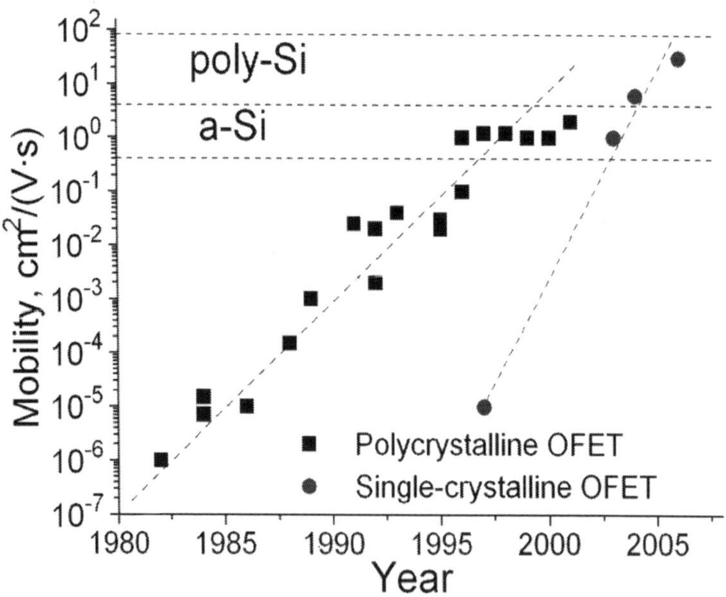

Figure: *Evolution of carrier mobility in organic field-effect transistor.*

The carrier transport in OFET is specific for two-dimensional (2D) carrier propagation through the device. Various experimental techniques were used for this study, such as Haynes - Shockley experiment on the transit times of injected carriers, time-of-flight (TOF) experiment for the determination of carrier mobility, pressure-wave propagation experiment for probing electric-field distribution in insulators, organic monolayer experiment for probing orientational dipolar changes, optical time-resolved second harmonic generation (TRM-SHG), etc. Whereas carriers propagate through polycrystalline OFETs in a diffusion-like (trap-limited) manner, they move through the conduction band in the best single-crystalline OFETs.

The most important parameter of OFET carrier transport is carrier mobility. Its evolution over the years of OFET research is shown in the graph for polycrystalline and single crystalline OFETs. The horizontal lines indicate the comparison guides to the main OFET competitors – amorphous (a-Si) and polycrystalline silicon. The graph reveals that the mobility in polycrystalline OFETs is comparable to that of a-Si whereas mobility in rubrene-based OFETs (20–40 $cm^2/(V \cdot s)$) approaches that of best poly-silicon devices.

Light-Emitting OFETs

Because an electric current flows through such a transistor, it can be used as a light-emitting device, thus integrating current modulation and light emission. In 2003, a German group reported the first organic light-emitting field-effect transistor (OLET). The device structure comprises interdigitated gold source- and drain electrodes and a polycrystalline tetracene thin film. Both, positive charges (holes) as well as negative charges (electrons) are injected from the gold contacts into this layer leading to electroluminescence from the tetracene.

Chemically Synthesized Semiconductor Nanowires for High Performance Electronics and Optoelectronics

A nanowire is a wire of dimensions of the order of a nanometre (10"9 metres). Alternatively, nanowires can be defined as structures that have a lateral size constrained to tens of nanometres or less and an unconstrained longitudinal size. At these scales, quantum mechanical effects are important — hence such wires are also known as "quantum wires".

Many different types of nanowires exist, including metallic (e.g., Ni, Pt, Au), semiconducting (e.g., InP, Si, GaN, etc.), and insulating (e.g., SiO2,TiO2).

Molecular nanowires are composed of repeating molecular units either organic or inorganic.

Carbon nanotubes have remarkable electronic, mechanical, and chemical properties. Depending on their specific diameter and the bonding arrangement of their carbon atoms, nanotubes exhibit either metallic or semiconducting behaviour. Electrical conduction within a perfect nanotube is ballistic (negligible scattering), with low thermal dissipation. As a result, a wire made from a nanotube, or a nanowire, can carry much more current than an ordinary metal wire of comparable size. At 1.4 nanometres in diameter, nanotubes are about a hundred times smaller than the gate width of silicon semiconductor devices. In addition to nanowires for conduction, transistors, diodes, and simple logic circuits have been demonstrated by combining metallic and semiconductor carbon nanotubes. Similarly, silicon nanowires have been used to build experimental devices, such as field-effect transistors, bipolar transistors, inverters, light-emitting diodes, sensors, and even simple memory. A major challenge for nanowire circuits, as for molecular electronics, is connecting and integrating these devices into a workable high-density architecture. Ideally, the structure would be grown and assembled in place. Crossbar architectures that combine the function of wires and devices are of particular interest.

Single-Electron Transistors

At nanoscale dimensions the energy required to add one additional electron to a "small island" (isolated physical region)—for example, through a tunneling barrier—becomes significant. This change in energy provides the basis for devising single-electron transistors. At low temperatures, where thermal fluctuations are small, various single-electron-device nanostructures are readily achievable, and extensive research has been carried out for structures with confined electron flow. However, room-temperature applications will require that sizes be reduced significantly, to the one-nanometre range, to achieve stable operation. For large-scale application with millions of devices, as found in current integrated circuits, the need for structures with very uniform size to maintain uniform device characteristics presents a significant challenge. Also, in this and many new nanodevices being explored, the lack of gain is a serious drawback limiting implementation in large-scale electronic circuits.

Spintronics

Spintronics refers to electronic devices that perform logic operations based on not just the electrical charge of carriers but also their spin. For example, information could be transported or stored through the spin-up or spin-down states of electrons. This is a new area of research, and issues include the injection of spin-polarized carriers, their transport, and their detection. The role of nanoscale structure and electronic properties of the ferromagnetic-semiconductor interface on the spin injection process, the growth of new ferromagnetic semiconductors with nanoscale control, and the possible use of nanostructured features to manipulate spin are all of interest.

Information Storage

Current approaches to information storage and retrieval include high-density, high-speed, solid-state electronic memories, as well as slower (but generally more spacious) magnetic and optical discs. As the minimum feature size for electronic processing approaches 100 nanometres, nanotechnology provides ways to decrease further the bit size of the stored information, thus increasing density and reducing interconnection distances for obtaining still-higher speeds. For example, the basis of the current generation of magnetic disks is the giant magnetoresistance effect. A magnetic read/write head stores bits of information by setting the direction of the magnetic field in nanometre-thick metallic layers that alternate between ferromagnetic and nonferromagnetic. Differences in spin-dependent scattering of electrons at the interface layers lead to resistance differences that can be read by the magnetic head. Mechanical properties, particularly tribology (friction and wear of moving surfaces), also play an important role in magnetic hard disk drives, since magnetic heads float only about 10 nanometres above spinning magnetic disks.

Another approach to information storage that is dependent on designing nanometre-thick magnetic layers is under commercial development. Known as magnetic random access memory (MRAM), a line of electrically switchable magnetic material is separated from a permanently magnetized layer by a nanoscale nonmagnetic interlayer. A resistance change that depends on the relative alignment of the fields is read electrically from a large array of wires through cross lines. MRAM will require a relatively small evolution from conventional semiconductor manufacturing, and it has the added benefit of producing nonvolatile memory (no power or batteries are needed to maintain stored memory states).

Still at an exploratory stage, studies of electrical conduction through molecules have generated interest in their possible use as memory. While still very speculative, molecular and nanowire approaches to memory are intriguing because of the small volume in which the bits of memory are stored and the effectiveness with which biological systems store large amounts of information.

2

Fast Flexible Electronics Made from Nanomembranes Derived from High Quality Wafers

In electronics, a wafer (also called a slice or substrate) is a thin slice of semiconductor material, such as a silicon crystal, used in the fabrication of integrated circuits and other microdevices. The wafer serves as the substrate for microelectronic devices built in and over the wafer and undergoes many microfabrication process steps such as doping or ion implantation, etching, deposition of various materials, and photolithographic patterning. Finally the individual microcircuits are separated (dicing) and packaged.

Several types of solar cell are also made from such wafers. On a *solar wafer* a solar cell (usually square) is made from the entire wafer. Wafers are formed of highly pure (99.9999999% purity), nearly defect-free single crystalline material. One process for forming crystalline wafers is known as Czochralski growth invented by the Polish chemist Jan Czochralski. In this process, a cylindrical ingot of high purity monocrystalline semiconductor, such as silicon or germanium, is formed by pulling a seed crystal from a 'melt'. Donor impurity atoms, such as boron or phosphorus in the case of silicon, can be added to the molten intrinsic material in precise amounts in order to dope the crystal, thus changing it into n-type or p-type extrinsic semiconductor.

The ingot is then sliced with a wafer saw (wire saw) and polished to form wafers. The size of wafers for photovoltaics is 100–200 mm square and the thickness is 200–300 μm. In the future, 160 μm will be the standard. Electronics use wafer sizes from 100–450 mm diameter.

Wafer Bonding

Wafer bonding is a packaging technology on wafer-level for the fabrication of microelectromechanical systems (MEMS), nanoelectromechanical systems (NEMS), microelectronics and optoelectronics, ensuring a mechanically stable and hermetically sealed encapsulation. The wafers' diameter range from 100 mm to 200 mm (4 inch to 8 inch) for MEMS/NEMS and up to 300 mm (12 inch) for the production of microelectronic devices. In microelectromechanical systems (MEMS) and nanoelectromechanical systems (NEMS), the package is an important part of the device. It enhances a proper function and protects the sensitive internal structures from environmental influences such as temperature, moisture, high pressure and oxidizing species. So, the long-term stability and reliability of the functional elements directly depend on the encapsulation process as well as a significant fraction on the overall device costs. In conclusion, the package has to fulfill the following requirements:

- protection against environmental influences
- heat dissipation
- integration of elements with different technologies
- compatibility to the surrounding periphery
- maintenance of energy and information flow

Techniques

The commonly used and developed bonding methods are as follows:

- Direct bonding
- Plasma activated bonding
- Anodic bonding
- Eutectic bonding
- Glass frit bonding
- Adhesive bonding
- Thermocompression bonding
- Reactive bonding
- Transient liquid phase diffusion bonding

Requirements

The bonding of wafers requires specific environmental conditions which can generally be defined as follows:

1. substrate surface
    - o flatness
    - o smoothness
    - o cleanliness

2. bonding environment
    - o bond temperature
    - o ambient pressure
    - o applied force
3. materials
    - o substrate materials
    - o intermediate layer materials

The actual bond is an interaction of all those conditions and requirements. Hence, the applied technology needs to be chosen in respect to the present substrate and defined specification like max. bearable temperature, mechanical pressure or desired gaseous atmosphere.

Evaluation

The bonded wafers are characterized in order to evaluate a technology's yield, bonding strength and level of hermeticity either for fabricated devices or for the purpose of process development. Therefore, several different approaches for the bond characterization have emerged. On the one hand non-destructive optical methods to find cracks or interfacial voids are used beside destructive techniques for the bond strength evaluation, like tensile or shear testing. On the other hand the unique properties of carefully chosen gases or the pressure depending vibration behaviour of micro resonators are exploited for hermeticity testing.

Flexible Electronics

Flexible electronics, also known as *flex circuits*, is a technology for assembling electronic circuits by mounting electronic devices on flexible plastic substrates, such as polyimide, PEEK or transparent conductive polyester film. Additionally, flex circuits can be screen printed silver circuits on polyester. Flexible electronic assemblies may be manufactured using identical components used for rigid printed circuit boards, allowing the board to conform to a desired shape, or to flex during its use. These flexible printed circuits (FPC) are made with a photolithographic technology. An alternative way of making flexible foil circuits or flexible flat cables (FFCs) is laminating very thin (0.07 mm) copper strips in between two layers of PET. These PET layers, typically 0.05 mm thick, are coated with an adhesive which is thermosetting, and will be activated during the lamination process. FPCs and FFCs have several advantages in many applications:

- Tightly assembled electronic packages, where electrical connections are required in 3 axes, such as cameras (static application).
- Electrical connections where the assembly is required to flex during its

normal use, such as folding cell phones (dynamic application).

- Electrical connections between sub-assemblies to replace wire harnesses, which are heavier and bulkier, such as in cars, rockets and satellites.
- Electrical connections where board thickness or space constraints are driving factors.

Figure: *An Olympus Stylus camera without the case, showing the flex circuit assembly.*

Applications

Flex circuits are often used as connectors in various applications where flexibility, space savings, or production constraints limit the serviceability of rigid circuit boards or hand wiring. A common application of flex circuits is in computer keyboards; most keyboards use flex circuits for the switch matrix.

In LCD fabrication, glass is used as a substrate. If thin flexible plastic or metal foil is used as the substrate instead, the entire system can be flexible, as the film deposited on top of the substrate is usually very thin, on the order of a few micrometres. Organic light-emitting diodes (OLEDs) are normally used instead of a back-light for flexible displays, making a flexible organic light-emitting diode display.

Most flexible circuits are passive wiring structures that are used to interconnect electronic components such as integrated circuits, resistor, capacitors and the like, however some are used only for making interconnections between other electronic assemblies either directly or by means of connectors.

In the automotive field, flexible circuits are used in instrument panels, under-hood controls, circuits to be concealed within the headliner of the cabin, and in ABS systems. In computer peripherals flexible circuits are used on the moving print head of printers, and to connect signals to the moving arm carrying the read/write heads of disk drives. Consumer electronics devices make use of flexible circuits in cameras, personal entertainment devices, calculators, or exercise monitors.

Flexible circuits are found in industrial and medical devices where many interconnections are required in a compact package. Cellular telephones are another widespread example of flexible circuits. Flexible solar cells have been developed for powering satellites. These cells are lightweight, can be rolled up for launch, and are easily deployable, making them a good match for the application. They can also be sewn into backpacks or outerwear.

Flexible circuits technology has a surprisingly long history. Patents issued at the turn of the 20th century show clear evidence that early researchers were envisioning ways of making flat conductors sandwiched between layers of insulating material to layout electrical circuits to serve in early telephony switching applications. One of the earliest descriptions of what could be called a flex circuit was unearthed by Dr Ken Gilleo and disclosed in an English patent by Albert Hansen in 1903 where Hansen described a construction consisting of flat metal conductors on paraffin coated paper. Thomas Edison's lab books from the same period also indicate that he was thinking to coat patterns cellulose gum applied to linen paper with graphite powder to create what would have clearly been flexible circuits, though there is no evidence that it was reduced to practice.

In the 1947 publication "Printed Circuit Techniques" by Cledo Brunetti and Roger W. Curtis (National Bureau of Standards Circular 468 first issued 15 November 1947) the a brief discussion of creating circuits on what would have been flexible insulating materials (e.g. paper) indicated that the idea was in place and in the 1950s Sanders Associates' inventors (Nashua, NH) Victor Dahlgren and company founder Royden Sanders made significant strides developing and patenting processes for printing and etching flat conductors on flexible base materials to replace wire harnesses. An advertisement from the 1950 placed by Photocircuits Corporation in New York demonstrated their active interest in flexible circuits also.

Today, flexible circuits which are also variously known around the world variously as *flexible printed wiring, flex print, flexi circuits,* are used many products. Large credit is due to the efforts of Japanese electronics packaging engineers who have found a countless new ways to employ flexible circuit technology. For the last decade, flexible circuits have remained one of the fastest growing of all interconnection product market segments. A more recent variation on flexible circuit technology is one called "flexible electronics" which commonly involves the integration of both active and passive functions in the processing.

Flexible Circuit Structures

There are a few basic constructions of flexible circuits but there is significant variation between the different types in terms of their construction. Following is a review of the most common types of flexible circuit constructions

Single-Sided Flex Circuits: Single-sided flexible circuits have a single conductor layer made of either a metal or conductive (metal filled) polymer on a flexible dielectric film. Component termination features are accessible only from one side. Holes may be formed in the base film to allow component leads to pass through for interconnection, normally by soldering. Single sided flex circuits can be fabricated with or without such protective coatings as cover layers or cover coats, however the use of a protective coating over circuits is the most common practice. The development of surface mounted devices on sputtered conductive films has enabled the production of transparent LED Films, which is used in LED Glass but also in flexible automotive lighting composites.

Double Access or Back Bared Flex Circuits: Double access flex, also known as back bared flex, are flexible circuits having a single conductor layer but which is processed so as to allow access to selected features of the conductor pattern from both sides. While this type of circuit has certain benefits, the specialised processing requirements for accessing the features limits its use.

Sculptured Flex Circuits: Sculptured flex circuits are a novel subset of normal flexible circuit structures. The manufacturing process involves a special flex circuit multi-step etching method which yields a flexible circuit having finished copper conductors wherein the thickness of the conductor differs at various places along their length. (i.e., the conductors are thin in flexible areas and thick at interconnection points.)

Double-Sided Flex Circuits: Double-sided flex circuits are flex circuits having two conductor layers. Theses flex circuits can be fabricated with or without plated through holes, though the plated through hole variation is much more common. When constructed without plated through holes and connection features are accessed from one side only, the circuit is defined as a "Type V (5)" according to military specifications. It is not a common practice but it is an option. Because of the plated through hole, terminations for electronic components are provided for on both sides of the circuit, thus allowing components to be placed on either side. Depending on design requirements, double-sided flex circuits can be fabricated with protective coverlayers on one, both or neither side of the completed circuit but are most commonly produced with the protective layer on both sides. One major advantage of this type of substrate is that it allows crossover connections to be made very easy. Many single sided circuits are built on a double sided substrate just because they have one of two crossover connections. An example of this use is the circuit connecting a mousepad to the motherboard of a laptop. All connections on that circuit are located on only one side of the substrate, except a very small crossover connection which uses the second side of the substrate.

Multilayer Flex Circuits: Flex circuits having three or more layers of conductors are known as multilayer flex circuits. Commonly the layers are interconnected by means of plated through holes, though this is not a requirement of the definition for it is possible to provide openings to access lower circuit level features. The layers of the multilayer flex circuit may or may not be continuously laminated together throughout the construction with the obvious exception of the areas occupied by plated through-holes. The practice of discontinuous lamination is common in cases where maximum flexibility is required. This is accomplished by leaving unbonded the areas where flexing or bending is to occur.

Rigid-Flex Circuits: Rigid-flex circuits are a hybrid construction flex circuit consisting of rigid and flexible substrates which are laminated together into a single structure. Rigid-flex circuits should not be confused with rigidized flex constructions, which are simply flex circuits to which a stiffener is attached to support the weight of the electronic components locally. A rigidized or stiffened flex circuit can have one or more conductor layers. Thus while the two terms may sound similar, they represent products that are quite different.

The layers of a rigid flex are also normally electrically interconnected by means of plated through holes. Over the years, rigid-flex circuits have enjoyed tremendous popularity among military product designer, however the technology has found increased use in commercial products. While often considered a speciality product for low volume applications because of the challenges, an impressive effort to use the technology was made by Compaq computer in the production of boards for a laptop computer in the 1990s. While the computer's main rigid-flex PCBA did not flex during use, subsequent designs by Compaq utilised rigid-flex circuits for the hinged display cable, passing 10s of 1000s of flexures during testing. By 2013, the use of rigid-flex circuits in consumer laptop computers is now common.

Rigid-flex boards are normally multilayer structures; however, two metal layer constructions are sometimes used.

Polymer Thick Film Flex Circuits: Polymer thick film (PTF) flex circuits are true printed circuits in that the conductors are actually printed onto a polymer base film. They are typically single conductor layer structures, however two or more metal layers can be printed sequentially with insulating layers printed between printed conductor layers. While lower in conductor conductivity and thus not suitable for all applications, PTF circuits have successfully served in a wide range of low-power applications at slightly higher voltages. Keyboards are a common application, however, there are a wide range of potential applications for this cost-effective approach to flex circuit manufacture.

Flexible Circuit Materials

Each element of the flex circuit construction must be able to consistently meet the demands placed upon it for the life of the product. In addition, the material must work reliably in concert with the other elements of the flexible circuit construction to assure ease of manufacture and reliability. Following are brief descriptions of the basic elements of flex circuit construction and their functions.

Base Material: The base material is the flexible polymer film which provides the foundation for the laminate. Under normal circumstances, the flex circuit base material provides most primary physical and electrical properties of the flexible circuit. In the case of adhesiveless circuit constructions, the base material provides all of the characteristic properties. While a wide range of thickness is possible, most flexible films are provided in a narrow range of relatively thin dimension from 12 μm to 125 μm (1/2 mil to 5 mils) but thinner and thicker material are possible. Thinner materials are of course more flexible and for most material, stiffness increase is proportional to the cube of thickness. Thus for example, means that if the thickness is doubled, the material becomes eight times stiffer and will only deflect 1/8 as much under the same load. There are a number of different materials used as base films including: polyester (PET), polyimide (PI), polyethylene napthalate (PEN), Polyetherimide (PEI), along with various fluropolymers (FEP) and copolymers Polyimide films are most prevalent owing to their blend of advantageous properties electrical, mechanical, chemical and thermal.

Bonding Adhesive: Adhesives are used as the bonding medium for creating a laminate. When it comes to temperature resistance, the adhesive is also typically is the performance limiting element of a laminate especially when polyimide is the base material. Because of the earlier difficulties associated with polyimide adhesives, many polyimide flex circuits presently employ adhesive systems of different polymer families. However some newer thermoplastic polyimide adhesives are making important in-roads. As with the base films, adhesives come in different thickness. Thickness selection is typically a function of the application. For example, different adhesive thickness is commonly used in the creation of cover layers in order to meet the fill demands of different copper foil thickness which may be encountered.

Metal Foil

A metal foil is most commonly used as the conductive element of a flexible laminate. The metal foil is the material from which the circuit paths are normally etched. A wide variety of metal foils of varying thickness are available from which to choose and create a flex circuit, however copper foils, serve the vast majority of all flexible circuit applications. Copper's excellent balance of cost and physical and electrical performance attributes make it an excellent choice. There are actually many different types of copper foil. The identifies eight different types of copper foil for printed circuits divided into two much broader categories, electrodeposited and wrought, each having four sub-types.) As a result, there are a number of different types of copper foil available for flex circuit applications to serve the varied purposes of different end products. With most copper foil, a thin surface treatment is commonly applied to one side of the foil to improve its adhesion to the base film. Copper foils are of two basic types: wrought (rolled) and electrodeposited and their properties are quite different. Rolled and annealed foils are the most common choice, however thinner films which are electroplated are becoming increasingly popular.

In certain non standard cases, the circuit manufacturer may be called upon to create a speciality laminate by using a specified alternative metal foil, such as a special copper alloy or other metal foil in the construction. This is accomplished by laminating the foil to a base film with or without an adhesive depending on the nature and properties of the base film.

Flexible Circuit Industry Standards and Specifications

Specifications are developed to provide a common ground of understanding of what a product should look like and how it should perform. Standards are developed directly by manufacturer's associations such as the Association Connecting Electronics Industries (IPC) and by users of flexible circuits.

Thin Films of Single Walled Carbon Nanotubes for Flexible Electronic Device Applications

Approaches for applying printing technologies to electronics manufacturing are attracting attention in recent years, giving rise to a new technology area called printable electronics.

Printable electronics can be characterized by simplified manufacturing process, shorter production time and reduced resource consumption compared to conventional device manufacturing technologies based on vacuum photolithography process. Moreover, manufacture of printable electronics can be carried out under low temperature and ordinary pressure conditions. These processes have scalability from ultrafine to large-area manufacturing and are expected to enable flexible electronics with diverse forms.

Manufacture of Thin Film Transistors (TFT) Using Printing Technology

The most challenging among all electronics manufacturing technologies utilising printing technology is the manufacture of the "basic semiconductor device," in other words the thin film transistor (TFT). If printing technology could enable thin film fabrication, patterning and lamination at room temperature, it will open thc door to various applications of printable electronics. Amorphous silicon and poly-silicon are currently the mainstream materials for TFT. TFTs made from these materials are indispensible as the driving element for flat panel displays, but they need to be processed under high temperature and vacuum conditions. Therefore, it is difficult with conventional technologies to manufacture TFTs on low-cost plastic film substrates. Various researches are being actively conducted to develop technologies for manufacturing TFTs on plastic films at low temperatures by using organic semiconductors, carbon nanotubes or oxide semiconductors as printing ink.

Among these potential technologies, carbon nanotubes are highly promising for application to flexible electronics including not only TFTs but also passive devices such as transparent conductive films and trace, because carbon nanotubes have a high mobility of 10-100cm2/Vs at room temperature in addition to mechanical stability (high mechanical strength and elasticity) and chemical stability (thermal and chemical resistance).

One challenge in using carbon nanotubes for TFT is the metal/semiconductor separation of carbon nanotubes. When single wall carbon nanotubes are grown using the CVD method, the resulting composition will be 70% semiconductor carbon nanotubes and 30% metal carbon nanotubes. Semiconductor carbon nanotubes should be used for application to TFTs and metal carbon nanotubes for electrodes.

Printed Carbon Nanotube Transistor Demonstrates World's Fastest Operation Speed

NEC and Technology Research Association for Single Wall Carbon Nanotubes (TASC) demonstrated operation speed of 500kHz with a printed carbon nanotube thin-film transistor (printed CNT-TFT) by reducing parasitic inductance when operated at high speed and improving output current by adopting high-purity carbon nanotube ink. This achievement opens the door to the application of printed CNT-TFTs as the control circuit of electronic devices (press release September 24, 2013).

Conventional TFTs could not be operated at a speed higher than several dozen kHz because electrodes would overlap with one another causing parasitic inductance to exceed the output current of the transistor. NEC and TASC succeeded in minimizing the overlap of electrodes by forming ultrafine electrodes of only 2μm in width using super inkjet printing technology, which enabled reduction of parasitic inductance to less than one tenth of common printed TNTs. This contributed to reducing the ratio of parasitic inductance with respect to total capacitance.

Proprietary technology for effective metal/semiconductor separation was developed by combining carrier-free electrophoresis with buoyancy. With this new technology, carbon nanotube purity was improved to 98% (from 95% conventionally) when conducting metal/semiconductor separation. A method to remove additives (CNT dispersion stabilizer) in the separation solution was also developed to reduce additive content in the CNT ink to a level below 1/50 compared to the conventional ink. This high-purity and low-additive CNT ink enables CNT thin films with higher conductivity than conventional ones, making it possible to produce printed CNT-TFTs with higher output current.

As a step towards developing a sheet device with the printed CNT TFT, prototypes of 16 x 16 transistor alley sheets have been created to verify the transistor characteristics (average mobility: 4.6cm2/Vs) with the overall elements on the sheet. In addition, pressure-sensitive conductive rubber and metal foil sheet were laminated on the CNT TFT to evaluate its application potential to sensor devices. It was confirmed that drain current was generated in response to applied pressure. Demonstration of high speed operation has opened up the possibility for control circuits using carbon nanotubes, leading our way to the realisation of flexible interface devices such as large-scale posters and sheet-type sensors.

Combining Semiconductor Polymer and Single Wall Carbon Nanotube

Toray developed a spray-coated carbon nanotube thin film transistor (coated CNT-TFT) demonstrating world's highest performance levels: mobility of 2.5cm2/Vs and on/off ratio of 106, by combining its proprietary semiconductor polymer and single wall carbon nanotubes (press release August 16, 2012).

When using single wall carbon nanotubes for thin film transistors, it is important that the carbon nanotubes are evenly dispersed between the electrodes to form the thin film. Since single wall carbon nanotubes have a strong tendency to bundle together, dispersants with strong electric insulation property are generally used to disperse the carbon nanotubes. However, the dispersants remaining between the single wall carbon nanotubes inhibit conductivity because of their insulation property, compromising TFT characteristics. To solve this problem, Toray discovered that the aggregation of carbon nanotubes can be prevented by applying a semiconductor polymer on the surface of single wall carbon nanotubes. This polymer, poly-3-hexyl thiophene (P3HT), prevents aggregation of the carbon nanotubes, like dispersants do, but it won't inhibit conductivity.

The semiconductor polymer was newly designed and synthesized by Toray so that it has a molecular structure suitable for wrapping around single wall carbon nanotubes of 1 nm in diameter. As a result, its adhesiveness to single wall carbon nanotubes has been improved compared to conventional P3HT. When this semiconductor polymer is applied by inkjet, nanotubes can form an evenly distributed single wall carbon nanotube thin film and at the same time realise high TFT characteristics.

The gate insulator film is an important component that is formed to insulate the gate electrodes which control the "on" and "off" of the transistor and the carbon nanotube thin film which conducts electricity. Toray, utilising its strengths in high polymer chemistry, developed an insulating material suitable for use in CNT-TFT. High mobility and low threshold voltage have been realised by incorporating compositional and molecular structural design to optimize the dispersion of the carbon nanotube thin film formed on the surface and to reduce threshold voltage, which is the challenge in CNT-TFT.

Demonstration study is underway for the application of coated CNT-TFT as transistors for displays. These efforts will be accelerated towards commercialization together with the development of suitable materials.

Field Electron Emission

Field emission (FE) (also known as field electron emission and electron field emission) is emission of electrons induced by an electrostatic field. The most common context is field emission from a solid surface into vacuum. However, field emission can take place from solid or liquid surfaces, into vacuum, air, a fluid, or any non-conducting or weakly conducting dielectric. The field-induced promotion of electrons from the valence to conduction band of semiconductors (the Zener effect) can also be regarded as a form of field emission. The terminology is historical because related phenomena of surface photoeffect, thermionic emission (or Richardson–Dushman effect) and "cold electronic emission", i.e. the emission of electrons in strong static (or quasi-static) electric fields, were discovered and studied independently from 1880s to 1930s. When field emission is used without qualifiers it typically means "cold emission".

Field emission in pure metals occurs in high electric fields: the gradients are typically higher than 1 gigavolt per metre and strongly dependent upon the work function. Electron sources based on field emission have a number of applications, but it is most commonly an undesirable primary source of vacuum breakdown and electrical discharge phenomena, which engineers work to prevent. Examples of applications for surface field emission include construction of bright electron sources for high-resolution electron microscopes or to discharge spacecraft from induced charges. Devices which eliminate induced charges are termed charge-neutralizers.

Field emission was explained by quantum tunneling of electrons in the late 1920s. This was one of the triumphs of the nascent quantum mechanics. The theory of field emission from bulk metals was proposed by Ralph H. Fowler and Lothar Wolfgang Nordheim. A family of approximate equations, "Fowler–Nordheim equations", is named after them. Strictly, Fowler–Nordheim equations apply only to field emission from bulk metals and (with suitable modification) to other bulk crystalline solids, but they are often used – as a rough approximation – to describe field emission from other materials.

In some respects, field electron emission is a paradigm example of what physicists mean by tunneling. Unfortunately, it is also a paradigm example of the intense mathematical difficulties that can arise. Simple solvable models of the tunneling barrier lead to equations (including the original 1928 Fowler–Nordheim-type equation) that get predictions of emission current density too low by a factor of 100 or more. If one inserts a more realistic barrier model into the simplest form of the Schrödinger equation, then an awkward mathematical problem arises over the resulting differential equation: it is known to be mathematically impossible in principle to solve this equation exactly in terms of the usual functions of mathematical physics, or in any simple way. To get even an approximate solution, it is necessary to use special approximate methods known in physics as "semi-classical" or "quasi-classical" methods. Worse, a mathematical error was made in the original application of these methods to field emission, and even the corrected theory that was put in place in the 1950s has been formally incomplete until very recently. A consequence of these (and other) difficulties has been a heritage of misunderstanding and disinformation that still persists in some current field emission research literature. This article tries to present a basic account of field emission "for the 21st century and beyond" that is free from these confusions.

Terminology and Conventions

Field electron emission, *field-induced electron emission*, *field emission* and *electron field emission* are general names for this experimental phenomenon and its theory. The first name is used here.

Fowler–Nordheim tunneling is the wave-mechanical tunneling of electrons through a rounded triangular barrier created at the surface of an electron conductor by applying a very high electric field. Individual electrons can escape by Fowler-Nordheim tunneling from many materials in various different circumstances.

Cold field electron emission (CFE) is the name given to a particular statistical emission regime, in which the electrons in the emitter are initially in internal thermodynamic equilibrium, and in which most emitted electrons escape by Fowler-Nordheim tunneling from electron states close to the emitter Fermi level. [By contrast, in the Schottky emission regime, most electrons escape over the top of a field-reduced barrier, from states well above the Fermi level.] Many solid and liquid materials can emit electrons in a CFE regime if an electric field of an appropriate size is applied.

Fowler–Nordheim-type equations are a family of approximate equations derived to describe CFE from the internal electron states in bulk metals. The different members of the family represent different degrees of approximation to reality. Approximate equations are necessary because, for physically realistic models of the tunneling barrier, it is mathematically impossible in principle to solve the Schrödinger equation exactly in any simple way. There is no theoretical reason to believe that Fowler-Nordheim-type equations validly describe field emission from materials other than bulk crystalline solids.

For metals, the CFE regime extends to well above room temperature. There are other electron emission regimes (such as "thermal electron emission" and "Schottky emission") that require significant external heating of the emitter. There are also emission regimes where the internal electrons are not in thermodynamic equilibrium and the emission current is, partly or completely, determined by the supply of electrons to the emitting region. A non-equilibrium emission process of this kind may be called field (electron) emission if most of the electrons escape by tunneling, but strictly it is not CFE, and is not accurately described by a Fowler-Nordheim-type equation.

Care is necessary because in some contexts (e.g. spacecraft engineering), the name "field emission" is applied to the field-induced emission of ions (field ion emission), rather than electrons, and because in some theoretical contexts "field emission" is used as a general name covering both field electron emission and field ion emission.

Historically, the phenomenon of field electron emission has been known by a variety of names, including "the aeona effect", "autoelectronic emission", "cold emission", "cold cathode emission", "field emission", "field electron emission" and "electron field emission".

Equations in this article are written using the International System of Quantities (ISQ). This is the modern (post-1970s) international system, based around the rationalized-metre-kilogram-second (rmks) system of equations, which is used to define SI units. Older field emission literature (and papers that directly copy equations from old literature) often write some equations using an older equation system that does not use the quantity ε_0. In this article, all such equations have been converted to modern international form. For clarity, this should always be done.

Since work function is normally given in electronvolts (eV), and it is often convenient to measure fields in volts per nanometre (V/nm), values of most universal constants are given here in units involving the eV, V and nm. Increasingly, this is normal practice in field emission research.

However, all equations here are ISQ-compatible equations and remain dimensionally consistent, as is required by the modern international system. To indicate their status, numerical values of universal constants are given to seven significant figures. Values are derived using the 2006 values of the fundamental constants.

Early History of Field Electron Emission

Field electron emission has a long, complicated and messy history. This section covers the early history, up to the derivation of the original Fowler–Nordheim-type equation in 1928.

In retrospect, it seems likely that the electrical discharges reported by Winkler in 1744 were started by CFE from his wire electrode. However, meaningful investigations had to wait until after J.J. Thomson's identification of the electron in 1897, and until after it was understood – from thermal emission and photo-emission work – that electrons could be emitted from inside metals (rather than from surface-adsorbed gas molecules), and that – in the absence of applied fields – electrons escaping from metals had to overcome a work function barrier.

It was suspected at least as early as 1913 that field-induced emission was a separate physical effect. However, only after vacuum and specimen cleaning techniques had significantly improved, did this become well established. Lilienfeld (who was primarily interested in electron sources for medical X-ray applications) published in 1922 the first clear account in English of the experimental phenomenology of the effect he had called "autoelectronic emission". He had worked on this topic, in Leipzig, since about 1910. Kleint describes this and other early work.

After 1922, experimental interest increased, particularly in the groups led by Millikan at the California Institute of Technology (Caltech) in Pasadena, California, and by Gossling at the General Electric Company in London. Attempts to understand autoelectronic emission included plotting experimental current-voltage (i - V) data in different ways, to look for a straight-line relationship. Current increased with voltage more rapidly than linearly, but plots of type (log(i) vs. V) were not straight. Schottky suggested in 1923 that the effect might be due to thermally induced emission over a field-reduced barrier. If so, then plots of type (log(i) vs. $V^{1/2}$) should be straight; but they were not. Nor is Schottky's explanation compatible with the experimental observation of only very weak temperature dependence in CFE – a point initially overlooked.

A breakthrough came when Lauritsen (and Oppenheimer independently) found that plots of type (log(i) vs. $1/V$) yielded good straight lines. This result, published by Millikan and Lauritsen in early 1928, was known to Fowler and Nordheim.

Oppenheimer had predicted that the field-induced tunneling of electrons from atoms (the effect now called field ionization) would have this $i(V)$ dependence, had found this dependence in the published experimental field emission results of Millikan and Eyring, and proposed that CFE was due to field-induced tunneling of electrons from atomic-like orbitals in surface metal atoms. An alternative Fowler–Nordheim theory explained both the Millikan-Lauritsen finding and the very weak dependence of current on temperature. Fowler–Nordheim theory predicted both to be consequences if CFE were due to field-induced tunneling from free-electron-type states in what we would now call a metal conduction band, with the electron states occupied in accordance with Fermi–Dirac statistics.

In fact, Oppenheimer (although right in principle about the theory of field ionization) had mathematical details of his theory seriously incorrect. There was also a small numerical error in the final equation given by Fowler–Nordheim theory for CFE current density: this was corrected in the 1929 paper of (Stern, Gossling & Fowler 1929).

Strictly, if the barrier field in Fowler-Nordheim 1928 theory is exactly proportional to the applied voltage, and if the emission area is independent of voltage, then the Fowler-Nordheim 1928 theory predicts that plots of the form ($\log(i/V^2)$ vs. $1/V$) should be exact straight lines. However, contemporary experimental techniques were not good enough to distinguish between the Fowler-Nordheim theoretical result and the Millikan-Lauritsen experimental result.

Thus, by 1928 basic physical understanding of the origin of CFE from bulk metals had been achieved, and the original Fowler-Nordheim-type equation had been derived.

The literature often presents Fowler-Nordheim work as a proof of the existence of electron tunneling, as predicted by wave-mechanics. Whilst this is correct, the validity of wave-mechanics was largely accepted by 1928. The more important role of the Fowler-Nordheim paper was that it was a convincing argument from experiment that Fermi–Dirac statistics applied to the behaviour of electrons in metals, as suggested by Sommerfeld in 1927. The success of Fowler–Nordheim theory did much to support the correctness of Sommerfeld's ideas, and greatly helped to establish modern electron band theory. In particular, the original Fowler-Nordheim-type equation was one of the first to incorporate the statistical-mechanical consequences of the existence of electron spin into the theory of an experimental condensed-matter effect. The Fowler-Nordheim paper also established the physical basis for a unified treatment of field-induced and thermally induced electron emission. Prior to 1928 it had been hypothesized that two types of electrons, "thermions" and "conduction electrons", existed in metals, and that thermally emitted electron currents were due to the emission of thermions, but that field-emitted currents were due to the emission of conduction electrons. The Fowler-Nordheim 1928 work suggested that thermions did not need to exist as a separate class of internal electrons: electrons could come from a single band occupied in accordance with Fermi–

Dirac statistics, but would be emitted in statistically different ways under different conditions of temperature and applied field.

The ideas of Oppenheimer, Fowler and Nordheim were also an important stimulus to the development, by Gurney and Condon, later in 1928, of the theory of the radioactive decay of nuclei (by alpha particle tunneling).

Practical Applications: Past and Present

Field Electron Microscopy and Related Basics: As already indicated, the early experimental work on field electron emission (1910–1920) was driven by Lilienfeld's desire to develop miniaturized X-ray tubes for medical applications. However, it was too early for this technology to succeed.

After Fowler-Nordheim theoretical work in 1928, a major advance came with the development in 1937 by Erwin W. Mueller of the spherical-geometry field electron microscope (FEM) (also called the "field emission microscope"). In this instrument, the electron emitter is a sharply pointed wire, of apex radius r. This is placed, in a vacuum enclosure, opposite an image detector (originally a phosphor screen), at a distance R from it. The microscope screen shows a projection image of the distribution of current-density J across the emitter apex, with magnification approximately (R/r), typically 10^5 to 10^6. In FEM studies the apex radius is typically 100 nm to 1 μm. The tip of the pointed wire, when referred to as a physical object, has been called a "field emitter", a "tip", or (recently) a "Mueller emitter".

When the emitter surface is clean, this FEM image is characteristic of: (a) the material from which the emitter is made: (b) the orientation of the material relative to the needle/wire axis; and (c) to some extent, the shape of the emitter endform. In the FEM image, dark areas correspond to regions where the local work function φ is relatively high and/or the local barrier field F is relatively low, so J is relatively low; the light areas correspond to regions where φ is relatively low and/or F is relatively high, so J is relatively high. This is as predicted by the exponent of Fowler-Nordheim-type equations.

The adsorption of layers of gas atoms (such as oxygen) onto the emitter surface, or part of it, can create surface electric dipoles that change the local work function of this part of the surface. This affects the FEM image; also, the change of work-function can be measured using a Fowler-Nordheim plot. Thus, the FEM became an early observational tool of surface science. For example, in the 1960s, FEM results contributed significantly to discussions on heterogeneous catalysis. FEM has also been used for studies of surface-atom diffusion. However, FEM has now been almost completely superseded by newer surface-science techniques.

A consequence of FEM development, and subsequent experimentation, was that it became possible to identify (from FEM image inspection) when an emitter was "clean", and hence exhibiting its clean-surface work-function as established by other techniques. This was important in experiments designed to test the validity of the standard Fowler-Nordheim-type equation. These experiments deduced a value of voltage-to-barrier-field conversion factor β from a Fowler-Nordheim plot, assuming the clean-surface φ–value for tungsten, and compared this with values derived from electron-microscope observations of emitter shape and electrostatic modelling. Agreement to within about 10% was achieved. Only very recently has it been possible to do the comparison the other way round, by bringing a well-prepared probe so close to a well-prepared surface that approximate parallel-plate geometry can be assumed and the conversion factor can be taken as $1/W$, where W is the measured probe-to emitter separation. Analysis of the resulting Fowler-Nordheim plot yields a work-function value close to the independently known work-function of the emitter.

Field Electron Spectroscopy (Electron Energy Analysis)

Energy distribution measurements of field-emitted electrons were first reported in 1939. In 1959 it was realised theoretically by Young, and confirmed experimentally by Young and Mueller that the quantity measured in spherical geometry was the distribution of the total energy of the emitted electron (its "total energy distribution"). This is because, in spherical geometry, the electrons move in such a fashion that angular momentum about a point in the emitter is very nearly conserved. Hence any kinetic energy that, at emission, is in a direction parallel to the emitter surface gets converted into energy associated with the radial direction of motion. So what gets measured in an energy analyzer is the total energy at emission.

With the development of sensitive electron energy analyzers in the 1960s, it became possible to measure fine details of the total energy distribution. These reflect fine details of the surface physics, and the technique of Field Electron Spectroscopy flourished for a while, before being superseded by newer surface-science techniques.

Field Electron Emitters as Electron-Gun Sources

To achieve high-resolution in electron microscopes and other electron beam instruments (such as those used for electron beam lithography), it is helpful to start with an electron source that is small, optically bright and stable. Sources based on the geometry of a Mueller emitter qualify well on the first two criteria. The first electron microscope (EM) observation of an individual atom was made by Crewe, Wall and Langmore in 1970, using a scanning electron microscope equipped with an early field emission gun.

From the 1950s onwards, extensive effort has been devoted to the development of field emission sources for use in electron guns. [e.g., DD53]

Methods have been developed for generating on-axis beams, either by field-induced emitter build-up, or by selective deposition of a low-work-function adsorbate (usually Zirconium oxide - ZrO) into the flat apex of a (100) oriented Tungsten emitter.

Sources that operate at room temperature have the disadvantage that they rapidly become covered with adsorbate molecules that arrive from the vacuum system walls, and the emitter has to be cleaned from time to time by "flashing" to high temperature. Nowadays, it is more common to use Mueller-emitter-based sources that are operated at elevated temperatures, either in the Schottky emission regime or in the so-called temperature-field intermediate regime. Many modern high-resolution electron microscopes and electron beam instruments use some form of Mueller-emitter-based electron source. Currently, attempts are being made to develop carbon nanotubes (CNTs) as electron-gun field emission sources.

The use of field emission sources in electron optical instruments has involved the development of appropriate theories of charged particle optics, and the development of related modelling. Various shape models have been tried for Mueller emitters; the best seems to be the "Sphere on Orthogonal Cone" (SOC) model introduced by Dyke, Trolan. Dolan and Barnes in 1953. Important simulations, involving trajectory tracing using the SOC emitter model, were made by Wiesener and Everhart. Nowadays, the facility to simulate field emission from Mueller emitters is often incorporated into the commercial electron-optics programmes used to design electron beam instruments. The design of efficient modern field-emission electron guns requires highly specialised expertise.

Atomically Sharp Emitters

Nowadays it is possible to prepare very sharp emitters, including emitters that end in a single atom. In this case, electron emission comes from an area about twice the crystallographic size of a single atom. This was demonstrated by comparing FEM and field ion microscope (FIM) images of the emitter. Single-atom-apex Mueller emitters also have relevance to the scanning probe microscopy and helium scanning ion microscopy (He SIM). Techniques for preparing them have been under investigation for many years. A related important recent advance has been the development (for use in the He SIM) of an automated technique for restoring a three-atom ("trimer") apex to its original state, if the trimer breaks up.Large-area field emission sources: vacuum nanoelectronics.

Materials Aspects: Large-area field emission sources have been of interest since the 1970s. In these devices, a high density of individual field emission sites is created on a substrate (originally silicon). This research area became known, first as "vacuum microelectronics", now as "vacuum nanoelectronics". One of the original two device types, the "Spindt array", used silicon-integrated-circuit (IC) fabrication techniques to make regular arrays in which molybdenum cones were deposited in small cylindrical voids in an oxide film, with the void covered by a counterelectrode with a central circular aperture. This overall geometry has also been used with carbon nanotubes grown in the void.

The other original device type was the "Latham emitter". These were MIMIV (metal-insulator-metal-insulator-vacuum) – or, more generally, CDCDV (conductor-dielectric-conductor-dielectric-vacuum) – devices that contained conducting particulates in a dielectric film. The device field-emits because its microstructure/nanostructure has field-enhancing properties. This material had a potential production advantage, in that it could be deposited as an "ink", so IC fabrication techniques were not needed. However in practice, uniformly reliable devices proved difficult to fabricate.

Research advanced to look for other materials that could be deposited/grown as thin films with suitable field-enhancing properties. In a parallel-plate arrangement, the "macroscopic" field F_M between the plates is given by $F_M = V/W$, where W is the plate separation and V is the applied voltage. If a sharp object is created on one plate, then the local field F at its apex is greater than F_M and can be related to F_M by

$$F = \gamma F_M.$$

The parameter γ is called the "field enhancement factor" and is basically determined by the object's shape. Since field emission characteristics are determined by the local field F, then the higher the γ-value of the object, then the lower the value of F_M at which significant emission occurs. Hence, for a given value of W, the lower the applied voltage V at which significant emission occurs.

For a roughly ten year-period from the mid-1990s, there was great interest in field emission from plasma-deposited films of amorphous and "diamond-like" carbon. However, interest subsequently lessened, partly due to the arrival of CNT emitters, and partly because evidence emerged that the emission sites might be associated with particulate carbon objects created in an unknown way during the deposition process: this suggested that quality control of an industrial-scale production process might be problematic.

The introduction of CNT field emitters, both in "mat" form and in "grown array" forms, was a significant step forwards. Extensive research has been undertaken into both their physical characteristics and possible technological applications. For field emission, an advantage of CNTs is that, due to their shape, with its high aspect ratio, they are "natural field-enhancing objects". Recently, it has been shown that CNTs exposed to microwave fields emit infrared, visible, and ultra-violet radiation as the result of field emission-induced luminescence.

Furthermore, it has been shown that the intensity of the radiation emitted by the CNTs during microwave irradiation increases over time as the CNTs are heated and tube ends open (resulting in a change in γ).

In recent years there has also been massive growth in interest in the development of other forms of thin-film emitter, both those based on other carbon forms (such as "carbon nanowalls") and on various forms of wide-band-gap semiconductor. A particular aim is to develop "high-γ" nanostructures with a sufficiently high density of individual emission sites.

Common problems with all field emission devices, particularly those that operate in "industrial vacuum conditions" is that the emission performance can be degraded by the adsorption of gas atoms arriving from elsewhere in the system, and the emitter shape can be in principle be modified deleteriously by a variety of unwanted subsidiary processes, such as bombardment by ions created by the impact of emitted electrons onto gas-phase atoms and/or onto the surface of counter-electrodes. Thus, an important industrial requirement is "robustness in poor vacuum conditions"; this needs to be taken into account in research on new emitter materials. At the time of writing, the most promising forms of large-area field emission source (certainly in terms of achieved average emission current density) seem to be Spindt arrays and the various forms of source based on CNTs.

Applications

The development of large-area field emission sources was originally driven by the wish to create new, more efficient, forms of electronic information display. These are known as "field emission displays" or "nano-emissive displays". Although several prototypes have been demonstrated, the development of such displays into reliable commercial products has been hindered by a variety of industrial production problems not directly related to the source characteristics [En08]. Other proposed applications of large-area field emission sources include microwave generation, space-vehicle neutralization, X-ray generation, and (for array sources) multiple e-beam lithography. There are also recent attempts to develop large-area emitters on flexible substrates, in line with wider trends towards "plastic electronics".

The development of such applications is the mission of vacuum nanoelectronics. However, field emitters work best in conditions of good ultrahigh vacuum. Their most successful applications to date (FEM, FES and EM guns) have occurred in these conditions. The sad fact remains that field emitters and industrial vacuum conditions do not go well together, and the related problems of reliably ensuring good "vacuum robustness" of field emission sources used in such conditions still await better solutions (probably cleverer materials solutions) than we currently have.

Vacuum Breakdown and Electrical Discharge Phenomena

As already indicated, it is now thought that the earliest manifestations of field electron emission were the electrical discharges it caused. After Fowler-Nordheim work, it was understood that CFE was one of the possible primary underlying causes of vacuum breakdown and electrical discharge phenomena. (The detailed mechanisms and pathways involved can be very complicated, and there is no single universal cause). Where vacuum breakdown is known to be caused by electron emission from a cathode, then the original thinking was that the mechanism was CFE from small conducting needle-like surface protrusions. Procedures were (and are) used to round and smooth the surfaces of electrodes that might generate unwanted field electron emission currents. However the work of Latham and others showed that emission could also be associated with the presence of semiconducting inclusions in smooth surfaces. The physics of how the emission is generated is still not fully understood, but suspicion exists that so-called "triple-junction effects" may be involved. Further information may be found in Latham's book and in the on-line bibliography.

Internal Electron Transfer in Electronic Devices

In some electronic devices, electron transfer from one material to another, or (in the case of sloping bands) from one band to another ("Zener tunneling"), takes place by a field-induced tunneling process that can be regarded as a form of Fowler-Nordheim tunneling. For example, Rhoderick's book discusses the theory relevant to metal-semiconductor contacts.

Fowler–Nordheim Tunneling

The next part of this article deals with the basic theory of cold field electron emission from bulk metals. This is best treated in four main stages, involving theory associated with: (1) derivation of a formula for "escape probability", by considering electron tunneling through a rounded triangular barrier; (2) an integration over internal electron states to obtain the "total energy distribution"; (3) a second integration, to obtain the emission current density as a function of local barrier field and local work function; (4) conversion of this to a formula for current as a function of applied voltage. The modified equations needed for large-area emitters, and issues of experimental data analysis, are dealt with separately.

Fowler–Nordheim tunneling is the wave-mechanical tunneling of an electron through an exact or rounded triangular barrier. Two basic situations are recognised: (1) when the electron is initially in a localized state; (2) when the electron is initially not strongly localized, and is best represented by a travelling wave. Emission from a bulk metal conduction band is a situation of the second type, and discussion here relates to this case. It is also assumed that the barrier is one-dimensional (i.e., has no lateral structure), and has no fine-scale structure that causes "scattering" or "resonance" effects. To keep this explanation of Fowler-Nordheim tunneling relatively simple, these assumptions are needed; but the atomic structure of matter is in effect being disregarded.

Motive Energy

For an electron, the one-dimensional Schrödinger equation can be written in the form

$$\frac{\hbar^2}{2m}\frac{d^2\Psi(x)}{dx^2} = [U(x) - E_n]\Psi(x) = M(x)\Psi(x), \tag{2}$$

where $\psi(x)$ is the electron wave-function, expressed as a function of distance x measured from the emitter's electrical surface, ' is the reduced Planck constant, m is the electron mass, $U(x)$ is the electron potential energy, E_n is the total electron energy associated with motion in the x-direction, and $M(x) = [U(x) - E_n]$ is called the electron motive energy. $M(x)$ can be interpreted as the negative of the electron kinetic energy associated with the motion of a hypothetical classical point electron in the x-direction, and is positive in the barrier.

The shape of a tunneling barrier is determined by how $M(x)$ varies with position in the region where $M(x) > 0$. Two models have special status in field emission theory: the *exact triangular (ET) barrier* and the *Schottky–Nordheim (SN) barrier*. These are given by equations (2) and (3), respectively:

$$M^{ET}(x) = h - eFx \tag{3}$$

$$M^{SN}(x) = h - eFx - e^2/(16\pi\varepsilon_0 x), \tag{4}$$

Here h is the zero-field height (or *unreduced height*) of the barrier, e is the elementary positive charge, F is the barrier field, and ε_0 is the electric constant. By convention, F is taken as positive, even though the classical electrostatic field would be negative. The SN equation uses the classical image potential energy to represent the physical effect "correlation and exchange".

Escape Probability

For an electron approaching a given barrier from the inside, the *probability of escape* (or "transmission coefficient" or "penetration coefficient") is a function of h and F, and is denoted by $D(h,F)$. The primary aim of tunneling theory is to calculate $D(h,F)$. For physically realistic barrier models, such as the Schottky-Nordheim barrier, the Schrödinger equation cannot be solved exactly in any simple way. The following so-called "semi-classical" approach can be used. A parameter $G(h,F)$ can be defined by the JWKB (Jeffreys-Wentzel-Kramers-Brillouin) integral:

$$G(h,F) = g\int M^{1/2}dx, \tag{4}$$

where the integral is taken across the barrier (i.e., across the region where $M > 0$), and the parameter g is a universal constant given by

$$g = 2\sqrt{2m} / \hbar \approx 10.24624 \text{ eV}^{-1/2} \text{ nm}^{-1}. \tag{5}$$

Forbes has re-arranged a result proved by Fröman and Fröman, to show that, formally – in a one-dimensional treatment – the exact solution for D can be written

$$D = \frac{P\text{e}^{-G}}{1 + P\text{e}^{-G}}, \tag{6}$$

where the *tunneling pre-factor* P can in principle be evaluated by complicated iterative integrations along a path in complex space. In the CFE regime we have (by definition) $G \gg 1$. Also, for simple models $P \approx 1$. So eq. (6) reduces to the so-called simple JWKB formula:

$$D \approx P\text{e}^{-G} \approx \text{e}^{-G}. \tag{7}$$

For the exact triangular barrier, putting eq. (2) into eq. (4) yields $G^{\text{ET}} = bh^{3/2}/F$, where

$$b = \frac{2g}{3e} = \frac{4\sqrt{2m}}{3e\hbar} \approx 6.830890 \text{ eV}^{-3/2} \text{ V nm}^{-1}. \tag{8}$$

This parameter b is a universal constant sometimes called the *second Fowler–Nordheim constant.* For barriers of other shapes, we write

$$G(h, F) = \nu(h, F) G^{\text{ET}} = \nu(h, F) b h^{3/2} / F, \tag{9}$$

where $\nu(h,F)$ is a correction factor that in general has to be determined by numerical integration, using eq. (4).

Correction Factor for the Schottky-Nordheim Barrier

The Schottky-Nordheim barrier, which is the barrier model used in deriving the standard Fowler-Nordheim-type equation, is a special case. In this case, it is known that the correction factor ν is a function of a single variable f_h, defined by $f_h = F/F_h$, where F_h is the field necessary to reduce the height of a Schottky–Nordheim barrier from h to 0. This field is given by

$$F_h = (4\pi\epsilon_0 / e^3) h^2 = (0.6944617 \text{ V nm}^{-1})(h / \text{eV})^2. \tag{10}$$

The parameter f_h runs from 0 to 1, and may be called the *scaled barrier field,* for a Schottky-Nordheim barrier of zero-field height h.

For the Schottky–Nordheim barrier, $\nu(h,F)$ is given by the particular value $\nu(f_h)$ of a function $\nu(\ell 2f)$. The latter is a function of mathematical physics in its own right and has been called the *principal Schottky–Nordheim barrier function.* An explicit series expansion for $\nu(\ell 2f)$ is derived in a 2008 paper by J. Deane. The following good simple approximation for $\nu(f_h)$ has been found:

$$\nu(f_h) \approx 1 - f_h + \tfrac{1}{6} f_h \ln f_h \tag{11}$$

Decay Width

The *decay width* (in energy), d_h, measures how fast the escape probability D decreases as the barrier height h increases; d_h is defined by:

$$\frac{1}{d_h} = -\frac{\mathrm{d}(\ln D)}{\mathrm{d}h}. \tag{12}$$

When h increases by d_h then the escape probability D decreases by a factor close to e ($\approx$ 2.718282). For an elementary model, based on the exact triangular barrier, where we put $\nu = 1$ and $P \approx 1$, we get

$$d_h^{(\mathrm{el})} = \frac{2F}{3b\sqrt{h}} = \frac{eF}{g\sqrt{h}}.$$

The decay width d_h derived from the more general expression (12) differs from this by a "decay-width correction factor" λ_d, so:

$$d_h = \lambda_d d_h^{(\mathrm{el})} = \frac{\lambda_d eF}{g\sqrt{h}}. \tag{13}$$

Usually, the correction factor can be approximated as unity.

The decay-width d_F for a barrier with h equal to the local work-function φ is of special interest. Numerically this is given by:

$$d_\mathrm{F} = \frac{\lambda_d eF}{g\sqrt{\phi}} \approx \frac{eF}{g\sqrt{\phi}} \approx 0.09759678\ \mathrm{eV}\cdot\sqrt{\frac{1\ \mathrm{eV}}{\phi}}\cdot\frac{F}{1\ \mathrm{V\ nm^{-1}}}. \tag{14}$$

For metals, the value of d_F is typically of order 0.2 eV, but varies with barrier-field F.

Comments

A historical note is necessary. The idea that the Schottky-Nordheim barrier needed a correction factor, as in eq. (9), was introduced by Nordheim in 1928, but his mathematical analysis of the factor was incorrect. A new (correct) function was introduced by Burgess, Kroemer and Houston in 1953, and its mathematics was developed further by Murphy and Good in 1956. This corrected function, sometimes known as a "special field emission elliptic function", was expressed as a function of a mathematical variable y known as the "Nordheim parameter".

Only recently (2006 to 2008) has it been realised that, mathematically, it is much better to use the variable –*!2* (= y^2). And only recently has it been possible to complete the definition of ν(*ℓ 2*) by developing and proving the validity of an exact series expansion for this function (by starting from known special-case solutions of the Gauss hypergeometric differential equation). Also, approximation (11) has been found only recently. Approximation (11) outperforms, and will presumably eventually displace, all older approximations of equivalent complexity. These recent developments, and their implications, will probably have a significant impact on field emission research in due course.

The following summary brings these results together. For tunneling well below the top of a well-behaved barrier of reasonable height, the escape probability $D(h,F)$ is given formally by:

$$D(h,F) \approx P\exp\left[-\frac{\nu(h,F)bh^{3/2}}{F}\right], \tag{15}$$

where $\nu(h,F)$ is a correction factor that in general has to be found by numerical integration. For the special case of a Schottky-Nordheim barrier, an analytical result exists and $\nu(h,F)$ is given by $\nu(f_h)$, as discussed above; approximation (11) for $\nu(f_h)$ is more than sufficient for all technological purposes. The pre-factor P is also in principle a function of h and (maybe) F, but for the simple physical models discussed here it is usually satisfactory to make the approximation $P = 1$. The exact triangular barrier is a special case where the Schrödinger equation can be solved exactly, as was done by Fowler and Nordheim; for this physically unrealistic case, $\nu(f_h) = 1$, and an analytical approximation for P exists.

The approach described here was originally developed to describe Fowler–Nordheim tunneling from smooth, classically flat, planar emitting surfaces. It is adequate for smooth, classical curved surfaces of radii down to about 10 to 20 nm. It can be adapted to surfaces of sharper radius, but quantities such as ν and D then become significant functions of the parameter(s) used to describe the surface curvature. When the emitter is so sharp that atomic-level detail cannot be neglected, and/or the tunneling barrier is thicker than the emitter-apex dimensions, then a more sophisticated approach is desirable.

As noted at the beginning, the effects of the atomic structure of materials are disregarded in the relatively simple treatments of field electron emission discussed here. Taking atomic structure properly into account is a very difficult problem, and only limited progress has been made. However, it seems probable that the main influences on the theory of Fowler-Nordheim tunneling will (in effect) be to change the values of P and ν in eq. (15), by amounts that cannot easily be estimated at present. All these remarks apply in principle to Fowler Nordheim tunneling from any conductor where (before tunneling) the electrons may be treated as in travelling-wave states. The approach may be adapted to apply (approximately) to situations where the electrons are initially in localized states at or very close inside the emitting surface, but this is beyond the scope of this article.

Total-Energy Distribution

The energy distribution of the emitted electrons is important both for scientific experiments that use the emitted electron energy distribution to probe aspects of the emitter surface physics and for the field emission sources used in electron beam instruments such as electron microscopes. In the latter case, the "width" (in energy) of the distribution influences how finely the beam can be focused.

The theoretical explanation here follows the approach of Forbes. If ε denotes the total electron energy relative to the emitter Fermi level, and K_p denotes the kinetic energy of the electron parallel to the emitter surface, then the electron's *normal energy* ε_n (sometimes called its "forwards energy") is defined by

$$\epsilon_n = \epsilon - K_p \tag{16}$$

Two types of theoretical energy distribution are recognised: the *normal-energy distribution* (NED), which shows how the energy ε_n is distributed immediately after emission (i.e., immediately outside the tunneling barrier); and the *total-energy distribution*, which shows how the total energy ε is distributed. When the emitter Fermi level is used as the reference zero level, both ε and ε_n can be either positive or negative.

Energy analysis experiments have been made on field emitters since the 1930s. However, only in the late 1950s was it realised (by Young and Mueller[,YM58]) that these experiments always measured the total energy distribution, which is now usually denoted by $j(\varepsilon)$. This is also true (or nearly true) when the emission comes from a small field enhancing protrusion on an otherwise flat surface. To see how the total energy distribution can be calculated within the framework of a Sommerfeld free-electron-type model, look at the *P-T energy-space diagram* (P-T="parallel-total").

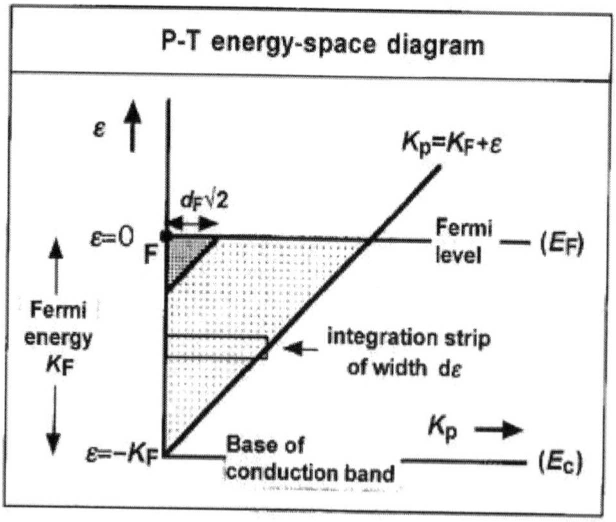

Figure: *P-T energy-space diagram, showing the region in P-T energy space where travelling-wave electron states exist.*

This shows the "parallel kinetic energy" K_p on the horizontal axis and the total energy ε on the vertical axis. An electron inside the bulk metal usually has values of K_p and ε that lie within the lightly shaded area. It can be shown that each element $\mathrm{d}\varepsilon\mathrm{d}K_p$ of this energy space makes a contribution $z_S f_{FD}\mathrm{d}\epsilon\mathrm{d}K_p$ to the electron current density incident on the inside of the emitter boundary. Here, z_S is the universal constant (called here the *Sommerfeld supply density*):

$$z_S = 4\pi em / h_P^3 = 1.618311 \times 10^{14}\,\mathrm{Am^{-2}\,eV^{-2}}, \tag{17}$$

and f_{FD} is the Fermi–Dirac distribution function:

$$f_{FD}(\epsilon) = 1/[1 + \exp(\epsilon / k_B T)], \tag{18}$$

where T is thermodynamic temperature and k_B is Boltzmann's constant.

This element of incident current density sees a barrier of height h given by:

$$h = \phi - \epsilon + K_p \text{...........}(19a)$$

The corresponding escape probability is $D(h,F)$: this may be expanded (approximately) in the form

$$D(h,F) \approx D_F \exp(\epsilon / d_F) \exp(-K_p / d_F) \text{..........}(19b),$$

where D_F is the escape probability for a barrier of unreduced height equal to the local work-function φ. Hence, the element $\mathrm{d}\varepsilon\mathrm{d}K_p$ makes a contribution $z_S f_{FD} D\mathrm{d}\epsilon\mathrm{d}K_p$ to the emission current density, and the total contribution made by incident electrons with energies in the elementary range dε is thus

$$j(\epsilon)\mathrm{d}\epsilon = z_S f_{FD}\left[\int D\mathrm{d}K_p\right]\mathrm{d}\epsilon = z_S f_{FD} D_F \exp(\epsilon / d_F)\left[\int_0^\infty \exp(-K_p / d_F)\,\mathrm{d}K_p\right]\mathrm{d}\epsilon, \tag{20}$$

where the integral is in principle taken along the strip shown in the diagram, but can in practice be extended to ∞ when the decay-width d_F is very much less than the Fermi energy K_F (which is always the case for a metal). The outcome of the integration can be written:

$$j(\epsilon) = z_S d_F D_F f_{FD}(\epsilon)\exp(\epsilon / d_F) = j_F f_{FD}(\epsilon)\exp(\epsilon / d_F), \tag{21}$$

where d_F and D_F are values appropriate to a barrier of unreduced height h equal to the local work function φ, and $j_F [= z_S d_F D_F]$ is defined by this equation.

For a given emitter, with a given field applied to it, j_F is independent of F, so eq. (21) shows that the shape of the distribution (as ε increases from a negative value well below the Fermi level) is a rising exponential, multiplied by the FD distribution function. This generates the familiar distribution shape first predicted by Young. At low temperatures, $f_{FD}(\epsilon)$ goes sharply from 1 to 0 in the vicinity of the Fermi level, and the FWHM of the distribution is given by:

$$\text{FWHM} = d_F \ln(2) \approx 0.693 d_F \tag{22}$$

The fact that experimental CFE total energy distributions have this basic shape is a good experimental confirmation that electrons in metals obey Fermi–Dirac statistics.

Equations for Cold Field Electron Emission (CFE)

Fowler–Nordheim-Type Equations (J-F forms)

Fowler–Nordheim-type equations, in the *J-F* form, are (approximate) theoretical equations derived to describe the local current density J emitted from the internal electron states in the conduction band of a bulk metal. The *emission current density* (ECD) J for some small uniform region of an emitting surface is usually expressed as a function $J(\varphi,F)$ of the local work-function φ and the local barrier field F that characterize the small region. For sharply curved surfaces, J may also depend on the parameter(s) used to describe the surface curvature. Owing to the physical assumptions made in the original derivation, the term *Fowler-Nordheim-type equation* has long been used only for equations that describe the ECD at zero temperature. However, it is better to allow this name to include the slightly modified equations (discussed below) that are valid for finite temperatures within the CFE emission regime.

Zero-Temperature Form

Current density is best measured in A/m². The total current density emitted from a small uniform region can be obtained by integrating the total energy distribution $j(\varepsilon)$ with respect to total electron energy ε. At zero temperature, the Fermi–Dirac distribution function $f_{FD} = 1$ for $\varepsilon<0$, and $f_{FD} = 0$ for $\varepsilon>0$. So the ECD at 0 K, J_0, is given from eq. (18) by

$$J_0 = z_S d_F D_F \int_{-\infty}^{0} \exp(\epsilon / d_F)\, d\epsilon = z_S d_F^{\,2} D_F = Z_F D_F, \qquad (23)$$

where $Z_F\,[= z_S d_F^{\,2}]$ is the *effective supply for state F*, and is defined by this equation. Strictly, the lower limit of the integral should be $-K_F$, where K_F is the Fermi Energy; but if d_F is very much less than K_F (which is always the case for a metal) then no significant contribution to the integral comes from energies below K_F, and it can formally be extended to $-\infty$.

Result (23) can be given a simple and useful physical interpretation by referring to Figure. The electron state at point "F" on the diagram ("state F") is the "forwards moving state at the Fermi level" (i.e., it describes a Fermi-level electron moving normal to and towards the emitter surface). At 0 K, an electron in this state sees a barrier of unreduced height φ, and has an escape probability D_F that is higher than that for any other occupied electron state. So it is convenient to write J_0 as $Z_F D_F$, where the "effective supply" Z_F is the current density that would have to be carried by state F inside the metal if all of the emission came out of state F.

In practice, the current density mainly comes out of a group of states close in energy to state F, most of which lie within the heavily shaded area in the energy-space diagram. Since, for a free-electron model, the contribution to the current density is directly proportional to the area in energy space (with the Sommerfeld supply density z_S as the constant of proportionality), it is useful to think of the ECD as drawn from electron states in an area of size d_F^2 (measured in eV^2) in the energy-space diagram. That is, it is useful to think of the ECD as drawn from states in the heavily shaded area in Figure earlier. (This approximation gets slowly worse as temperature increases.)

Z_F can also be written in the form:

$$Z_F = z_S d_F^2 = \lambda_d^2 (z_S e^2 g^{-2}) \phi^{-1} F^2 = \lambda_d^2 a \phi^{-1} F^2, \tag{24}$$

where the universal constant a, sometimes called the *First Fowler–Nordheim Constant*, is given by

$$a = z_S e^2 g^{-2} = e^3 / 8\pi h_P \approx 1.541434 \times 10^{-6} \text{ A eV V}^{-2} \tag{25}$$

This shows clearly that the pre-exponential factor $a\ \varphi^{-1} F^2$, that appears in Fowler-Nordheim-type equations, relates to the effective supply of electrons to the emitter surface, in a free-electron model.

Non-Zero Temperatures

To obtain a result valid for non-zero temperature, we note from eq. (23) that $z_S d_F D_F = J_0/d_F$. So when eq. (21) is integrated at non-zero temperature, then – on making this substitution, and inserting the explicit form of the Fermi–Dirac distribution function – the ECD J can be written in the form:

$$J = J_0 \int_{-\infty}^{\infty} \frac{\exp(\epsilon / d_F)}{[1 + \exp[(\epsilon / d_F)(d_F / k_B T)]]} d(\epsilon / d_F) = \lambda_T J_0, \tag{26}$$

where λ_T is a temperature correction factor given by the integral. The integral can be transformed, by writing $w = d_F / k_B T$ and $x = \epsilon / d_F$, and then $u = \exp(x)$, into the standard result:

$$\int_{-\infty}^{\infty} [e^x / (1 + e^{wx})]\, dx = \int_0^{\infty} [u / (1 + wu)] du = (\pi / w) / \sin(\pi / w) \tag{27}$$

This is valid for $w>1$ (i.e., $d_F/k_B T > 1$). Hence – for temperatures such that $k_B T < d_F$:

$$\lambda_T = (\pi k_B T / d_F) / \sin(\pi k_B T / d_F) \approx 1 + (1/6)(\pi k_B T / d_F)^2, \tag{28}$$

where the expansion is valid only if $(\pi k_B T / d_F) << 1$. An example value (for $\varphi = 4.5$ eV, $F = 5$ V/nm, $T = 300$ K) is $\lambda_T = 1.024$. Normal thinking has been that, in the CFE regime, λ_T is always small in comparison with other uncertainties, and that it is usually unnecessary to explicitly include it in formulae for the current density at room temperature.

The emission regimes for metals are, in practice, defined, by the ranges of barrier field F and temperature T for which a given family of emission equations is mathematically adequate. When the barrier field F is high enough for the CFE regime to be operating for metal emission at 0 K, then the condition $k_B T < d_F$ provides a formal upper bound (in temperature) to the CFE emission regime. However, it has been argued that (due to approximations made elsewhere in the derivation) the condition $k_B T < 0.7 d_F$ is a better working limit: this corresponds to a λ_T-value of around 1.09, and (for the example case) an upper temperature limit on the CFE regime of around 1770 K. This limit is a function of barrier field.

Note that result (28) here applies for a barrier of any shape (though d_F will be different for different barriers).

Physically Complete Fowler-Nordheim-Type Equation

Result (23) also leads to some understanding of what happens when atomic-level effects are taken into account, and the band-structure is no longer free-electron like. Due to the presence of the atomic ion-cores, the surface barrier, and also the electron wave-functions at the surface, will be different. This will affect the values of the correction factor ν, the prefactor P, and (to a limited extent) the correction factor λ_d. These changes will, in turn, affect the values of the parameter D_F and (to a limited extent) the parameter d_F. For a real metal, the supply density will vary with position in energy space, and the value at point "F" may be different from the Sommerfeld supply density. We can take account of this effect by introducing an electronic-band-structure correction factor λ_B into eq. (23). Modinos has discussed how this factor might be calculated: he estimates that it is most likely to be between 0.1 and 1; it might lie outside these limits but is most unlikely to lie outside the range $0.01 < \lambda_B < 10$.

By defining an overall supply correction factor λ_Z equal to $\lambda_T \lambda_B \lambda_d^2$, and combining equations above, we reach the so-called physically complete Fowler-Nordheim-type equation:

$$J = \lambda_Z a \phi^{-1} F^2 P_F \exp[-\nu_F b \phi^{3/2} / F], \tag{29}$$

where ν_F [$= \nu_F(\varphi, F)$] is the exponent correction factor for a barrier of unreduced height φ. This is the most general equation of the Fowler–Nordheim type. Other equations in the family are obtained by substituting specific expressions for the three correction factors ν_F, P_F and λ_Z it contains. The so-called elementary Fowler-Nordheim-type equation, that appears in undergraduate textbook discussions of field emission, is obtained by putting $\lambda_Z \to 1$, $P_F \to 1$, $\to 1$; this does not yield good quantitative predictions because it makes the barrier stronger than it is in physical reality. The so-called standard Fowler-Nordheim-type equation, originally developed by Murphy and Good, and much used in past literature, is obtained by putting $\lambda_Z \to t_F^{-2}$, $P_F \to 1$, $\nu_F \to \upsilon_F$, where υ_F is $\upsilon(f)$, where f is the value of f_h obtained by putting $h=\varphi$, and t_F is a related parameter (of value close to unity).

Within the more complete theory described here, the factor t_F^{-2} is a component part of the correction factor λ_d^2 [see, and note that λ_d^2 is denoted by λ_D there]. There is no significant value in continuing the separate identification of t_F^{-2}. Probably, in the present state of knowledge, the best approximation for simple Fowler-Nordheim-type equation based modelling of CFE from metals is obtained by putting $\lambda_Z \to 1$, $P_F \to 1$, $\nu_F \to \upsilon(f)$. This re-generates the Fowler-Nordheim-type equation used by Dyke and Dolan in 1956, and can be called the "simplified standard Fowler-Nordheim-type equation".

Recommended form for Simple Fowler-Nordheim-Type Calculations

Explicitly, this recommended *simplified standard Fowler-Nordheim-type equation*, and associated formulae, are:

$$J = a\phi^{-1}F^2\exp[-v(f)\,b\phi^{3/2} / F], \tag{30a}$$

$$a \approx 1.541434\times10^{-6}\ \text{A eV V}^{-2}; \quad b \approx 6.830890\ \text{eV}^{-3/2}\ \text{V nm}^{-1}, \tag{30b}$$

$$v(f) \approx 1 - f + (1/6)f\ln f \tag{30c}$$

$$f = F / F_\phi = (e^3 / 4\pi\epsilon_0)(F / \phi^2) = (1.439964\ \text{eV}^2\ \text{V}^{-1}\ \text{nm})(F / \phi^2) \tag{30d}$$

where F_φ here is the field needed to reduce to zero a Schottky-Nordheim barrier of unreduced height equal to the local work-function φ, and f is the scaled barrier field for a Schottky-Nordheim barrier of unreduced height φ. [This quantity f could have been written more exactly as f_φ^{SN}, but it makes this Fowler-Nordheim-type equation look less cluttered if the convention is adopted that simple f means the quantity denoted by f_φ^{SN} in, eq. (2.16).] For the example case (φ= 4.5 eV, F= 5 V/nm), $f \approx$ 0.36 and $\upsilon(f) \approx$ 0.58; practical ranges for these parameters are discussed further in.

Note that the variable f (the scaled barrier field) is not the same as the variable y (the Nordheim parameter) extensively used in past field emission literature, and that " $\upsilon(f)$ " does NOT have the same mathematical meaning and values as the quantity " $\upsilon(y)$ " that appears in field emission literature. In the context of the revised theory described here, formulae for $\upsilon(y)$, and tables of values for $\upsilon(y)$ should be disregarded, or treated as values of $\upsilon(f^{1/2})$. If more exact values for $\upsilon(f)$ are required, then provides formulae that give values for $\upsilon(f)$ to an absolute mathematical accuracy of better than 8×10^{-10}. However, approximation formula (30c) above, which yields values correct to within an absolute mathematical accuracy of better 0.0025, should gives values sufficiently accurate for all technological purposes.

Comments

A historical note on methods of deriving Fowler-Nordheim-type equations is necessary. There are several possible approaches to deriving these equations, using free-electron theory. The approach used here was introduced by Forbes in 2004 and may be described as "integrating via the total energy distribution, using the parallel kinetic energy K_p as the first variable of integration". Basically, it is a free-electron equivalent of the Modinos procedure (in a more advanced quantum-mechanical treatment) of "integrating over the surface Brillouin zone". By contrast, the free-electron treatments of CFE by Young in 1959, Gadzuk and Plummer in 1973 and Modinos in 1984, also integrate via the total energy distribution, but use the normal energy ε_n (or a related quantity) as the first variable of integration.

There is also an older approach, based on a seminal paper by Nordheim in 1928, that formulates the problem differently and then uses first K_p and then ε_n (or a related quantity) as the variables of integration: this is known as "integrating via the normal-energy distribution". This approach continues to be used by some authors. Although it has some advantages, particularly when discussing resonance phenomena, it requires integration of the Fermi–Dirac distribution function in the first stage of integration: for non-free-electron-like electronic band-structures this can lead to very complex and error-prone mathematics (as in the work of Stratton on semiconductors). Further, integrating via the normal-energy distribution does not generate experimentally measured electron energy distributions.

In general, the approach used here seems easier to understand, and leads to simpler mathematics. It is also closer in principle to the more sophisticated approaches used when dealing with real bulk crystalline solids, where the first step is either to integrate contributions to the ECD over constant energy surfaces in a wave-vector space (k -space), or to integrate contributions over the relevant surface Brillouin zone. The Forbes approach is equivalent either to integrating over a spherical surface in k -space, using the variable K_p to define a ring-like integration element that has cylindrical symmetry about an axis in a direction normal to the emitting surface, or to integrating over an (extended) surface Brillouin zone using circular-ring elements.

CFE Theoretical Equations (i-V Forms)

The preceding section explains how to derive Fowler-Nordheim-type equations. Strictly, these equations apply only to CFE from bulk metals. The ideas in the following sections apply to CFE more generally, but eq. (30) will be used to illustrate them.

For CFE, basic theoretical treatments provide a relationship between the local emission current density J and the local barrier field F, at a local position on the emitting surface. Experiments measure the emission current i from some defined part of the emission surface, as a function of the voltage V applied to some counter-electrode. To relate these variables to J and F, auxiliary equations are used.

The *voltage-to-barrier-field conversion factor* β is defined by:

$$F = \beta V, \tag{31}$$

The value of F varies from position to position on an emitter surface, and the value of β varies correspondingly.

For a metal emitter, the β"value for a given position will be constant (independent of voltage) under the following conditions: (1) the apparatus is a "diode" arrangement, where the only electrodes present are the emitter and a set of "surroundings", all parts of which are at the same voltage; (2) no significant field-emitted vacuum space-charge (FEVSC) is present (this will be true except at very high emission current densities, around 10^9 A/m^2 or higher); (3) no significant "patch fields" exist, as a result of non-uniformities in local work-function (this is normally assumed to be true, but may not be in some circumstances). For non-metals, the physical effects called "field penetration" and "band bending" [M084] can make β a function of applied voltage, although – surprisingly – there are few studies of this effect.

The emission current density J varies from position to position across the emitter surface. The total emission current i from a defined part of the emitter is obtained by integrating J across this part. To obtain a simple equation for $i(V)$, the following procedure is used. A reference point "r" is selected within this part of the emitter surface (often the point at which the current density is highest), and the current density at this reference point is denoted by J_r. A parameter A_r, called the *notional emission area* (with respect to point "r"), is then defined by:

$$i = A_r J_r = \int J \mathrm{d}A, \tag{32}$$

where the integral is taken across the part of the emitter of interest.

This parameter A_r was introduced into CFE theory by Stern, Gossling and Fowler in 1929 (who called it a "weighted mean area"). For practical emitters, the emission current density used in Fowler-Nordheim-type equations is always the current density at some reference point (though this is usually not stated). Long-established convention denotes this reference current density by the simple symbol J, and the corresponding local field and conversion factor by the simple symbols F and β, without the subscript "r" used above; in what follows, this convention is used.

The notional emission area A_r will often be a function of the reference local field (and hence voltage), and in some circumstances might be a significant function of temperature.

Because A_r has a mathematical definition, it does not necessarily correspond to the area from which emission is observed to occur from a single-point emitter in a field electron (emission) microscope. With a large-area emitter, which contains many individual emission sites, A_r will nearly always be very very much less than the "macroscopic" geometrical area (A_M) of the emitter as observed visually.

Incorporating these auxiliary equations into eq. (30a) yields

$$i = A_r a\phi^{-1}\beta^2 V^2 \exp[-v(f)\, b\phi^{3/2} / \beta V], \tag{33}$$

This is the simplified standard Fowler-Nordheim-type equation, in *i-V* form. The corresponding "physically complete" equation is obtained by multiplying by $\lambda_Z P_F$.

Modified Equations for Large-Area Emitters

The equations in the preceding section apply to all field emitters operating in the CFE regime. However, further developments are useful for large-area emitters that contain many individual emission sites.

For such emitters, the notional emission area will nearly always be very very much less than the apparent "macroscopic" geometrical area (A_M) of the physical emitter as observed visually. A dimensionless parameter α_r, *the area efficiency of emission*, can be defined by

$$A_r = \alpha_r A_M \tag{34}$$

Also, a "macroscopic" (or "mean") emission current density J_M (averaged over the geometrical area A_M of the emitter) can be defined, and related to the reference current density J_r used above, by

$$J_M = i / A_M = \alpha_r (i / A_r) = \alpha_r J_r \tag{35}$$

This leads to the following "large-area versions" of the simplified standard Fowler-Nordheim-type equation:

$$J_M = \alpha_r a\phi^{-1} F^2 \exp[-v(f)\, b\phi^{3/2} / F], \tag{36}$$

$$i = \alpha_r A_M a\phi^{-1}\beta^2 V^2 \exp[-v(f)\, b\phi^{3/2} / \beta V], \tag{37}$$

Both these equations contain the area efficiency of emission α_r. For any given emitter this parameter has a value that is usually not well known. In general, α_r varies greatly as between different emitter materials, and as between different specimens of the same material prepared and processed in different ways. Values in the range 10^{-10} to 10^{-6} appear to be likely, and values outside this range may be possible.

The presence of α_r in eq. (36) accounts for the difference between the macroscopic current densities often cited in the literature (typically 10 A/m^2 for many forms of large-area emitter other than Spindt arrays) and the local current densities at the actual emission sites, which can vary widely but which are thought to be generally of the order of 10^9 A/m^2, or possibly slightly less.

A significant part of the technological literature on large-area emitters fails to make clear distinctions between local and macroscopic current densities, or between notional emission area A_r and macroscopic area A_M, and/or omits the parameter α_r from cited equations. Care is necessary in order to avoid errors of interpretation.

It is also sometimes convenient to split the conversion factor α_r into a "macroscopic part" that relates to the overall geometry of the emitter and its surroundings, and a "local part" that relates to the ability of the very-local structure of the emitter surface to enhance the electric field. This is usually done by defining a "macroscopic field" F_M that is the field that would be present at the emitting site in the absence of the local structure that causes enhancement. This field F_M is related to the applied voltage by a "voltage-to-macroscopic-field conversion factor" α_M defined by:

$$F_M = \beta_M V \tag{38}$$

In the common case of a system comprising two parallel plates, separated by a distance W, with emitting nanostructures created on one of them, $\alpha_M = 1/W$.

A "field enhancement factor" λ is then defined and related to the values of α_r and α_M by

$$\gamma = F_r / F_M = \beta_r / \beta_M \tag{39}$$

With eq. (31), this generates the following formulae:

$$F = \gamma F_M = \beta V \tag{41}$$

$$\beta = \beta_M \gamma \tag{42}$$

where, in accordance with the usual convention, the suffix "r" has now been dropped from parameters relating to the reference point. Formulae exist for the estimation of λ, using classical electrostatics, for a variety of emitter shapes, in particular the "hemisphere on a post".

Equation (40) implies that versions of Fowler-Nordheim-type equations can be written where either F or αV is everywhere replaced by γF_M. This is often done in technological applications where the primary interest is in the field enhancing properties of the local emitter nanostructure. However in some past work, failure to make a clear distinction between barrier field F and macroscopic field F_M has caused confusion or error.

More generally, the aims in technological development of large-area field emitters are to enhance the uniformity of emission by increasing the value of the area efficiency of emission α_r, and to reduce the "onset" voltage at which significant emission occurs, by increasing the value of α. Eq. (41) shows that this can be done in two ways: either by trying to develop "high-λ" nanostructures, or by changing the overall geometry of the system so that β_M is increased. Various trade-offs and constraints exist.

In practice, although the definition of macroscopic field used above is the commonest one, other (differently defined) types of macroscopic field and field enhancement factor are used in the literature, particularly in connection with the use of probes to investigate the *i-V* characteristics of individual emitters.

In technological contexts field emission data are often plotted using (a particular definition of) F_M or $1/F_M$ as the *x*-coordinate. However, for scientific analysis it usually better not to pre-manipulate the experimental data, but to plot the raw measured *i-V* data directly. Values of technological parameters such as (the various forms of) γ can then be obtained from the fitted parameters of the *i-V* data plot, using the relevant definitions.

Empirical CFE i-V *Equation*

At the present stage of CFE theory development, it is important to make a distinction between theoretical CFE equations and an empirical CFE equation. The former are derived from condensed matter physics (albeit in contexts where their detailed development is difficult). An empirical CFE equation, on the other hand, simply attempts to represent the actual experimental form of the dependence of current *i* on voltage *V*.

In thc 1920s, empirical equations were used to find the power of *V* that appeared in the exponent of a semi-logarithmic equation assumed to describe experimental CFE results. In 1928, theory and experiment were brought together to show that (except, possibly, for very sharp emitters) this power is V^{-1}. It has recently been suggested that CFE experiments should now be carried out to try to find the power (κ) of *V* in the pre-exponential of the following empirical CFE equation:

$$i = CV^{\kappa}\exp[-B/V], \tag{42}$$

where *B*, *C* and κ are treated as constants.

From eq. (42) it is readily shown that

$$-\mathrm{d}\ln i/\mathrm{d}(1/V) = \kappa V + B, \tag{43}$$

In the 1920s, experimental techniques could not distinguish between the results κ =0 (assumed by Millikan and Laurtisen) and κ=2 (predicted by the original Fowler-Nordheim-type equation). However, it should now be possible to make reasonably accurate measurements of dlni/d(1/V) (if necessary by using lock-in amplifier/phase-sensitive detection techniques and computer-controlled equipment), and to derive κ from the slope of an appropriate data plot.

Following the discovery of approximation (30b), it is now very clear that – even for CFE from bulk metals – the value κ=2 is not expected. This can be shown as follows. Using eq. (30c) above, a dimensionless parameter η may be defined by

$$\eta = b\phi^{3/2} / F_{\phi} = (be^{3} / 4\pi\epsilon_{0})\phi^{-1/2} \approx 9.836239\ (\mathrm{eV} / \phi)^{1/2} \tag{34}$$

For φ = 4.50 eV, this parameter has the value η = 4.64. Since $f = F/F_{\varphi}$ and *υ*(*f*) is given by eq (30b), the exponent in the simplified standard Fowler-Nordheim-type equation (30) can be written in an alternative form and then expanded as follows:

$$\exp[-v(f)\, b\phi^{3/2} / F] = \exp[-v(f)\, \eta / f] \approx e^{\eta} f^{-\eta/6} \exp[-\eta / f] = e^{\eta} f^{-\eta/6} \exp[-b\phi^{3/2} / F] \tag{35}$$

Provided that the conversion factor β is independent of voltage, the parameter *f* has the alternative definition $f = V/V_{\varphi}$, where V_{φ} is the voltage needed, in a particular experimental system, to reduce the height of a Schottky-Nordheim barrier from φ to zero. Thus, it is clear that the factor *υ*(*f*) in the *exponent* of the theoretical equation (30) gives rise to additional *V*-dependence in the *pre-exponential* of the empirical equation. Thus, (for effects due to the Schottky-Nordheim barrier, and for an emitter with φ=4.5 eV) we obtain the prediction:

$$\kappa \approx 2 - \eta / 6 = 2 - 0.77 = 1.23 \tag{46}$$

Since there may also be voltage dependence in other factors in a Fowler-Nordheim-type equation, in particular in the notional emission area A_r and in the local work-function, it is not necessarily expected that κ for CFE from a metal of local work-function 4.5 eV should have the value κ = 1.23, but there is certainly no reason to expect that it will have the original Fowler-Nordheim value κ = 2.

A first experimental test of this proposal has been carried out by Kirk, who used a slightly more complex form of data analysis to find a value 1.36 for his parameter κ. His parameter κ is very similar to, but not quite the same as, the parameter κ used here, but nevertheless his results do appear to confirm the potential usefulness of this form of analysis.

Use of the empirical CFE equation (42), and the measurement of κ, may be of particular use for non-metals. Strictly, Fowler-Nordheim-type equations apply only to emission from the conduction band of bulk crystalline solids. However, empirical equations of form (42) should apply to all materials (though, conceivably, modification might be needed for very sharp emitters). It seems very likely that one way in which CFE equations for newer materials may differ from Fowler-Nordheim-type equations is that these CFE equations may have a different power of *F* (or *V*) in their pre-exponentials. Measurements of κ might provide some experimental indication of this.

Fowler-Nordheim Plots and Millikan-Lauritsen Plots

The original theoretical equation derived by Fowler and Nordheim has, for the last 80 years, influenced the way that experimental CFE data has been plotted and analyzed. In the very widely used Fowler-Nordheim plot, as introduced by Stern *et al.* in 1929, the quantity $\ln\{i/V^2\}$ is plotted against $1/V$. The original thinking was that (as predicted by the original or the elementary Fowler-Nordheim-type equation) this would generate an exact straight line of slope S_{FN}. S_{FN} would be related to the parameters that appear in the exponent of a Fowler-Nordheim-type equation of i-V form by:

$$S_{FN} = -b\phi^{3/2}/\beta \tag{47}$$

Hence, knowledge of φ would allow β to be determined, or vice-versa.

[In principle, in system geometries where there is local field-enhancing nanostructure present, and the macroscopic conversion factor β_M can be determined, knowledge of β then allows the value of the emitter's effective field enhancement factor γ to be determined from the formula $\gamma = \beta/\beta_M$. In the common case of a film emitter generated on one plate of a two-plate arrangement with plate-separation W (so $\beta_M = 1/W$) then

$$\gamma = \beta W \tag{48}$$

Nowadays, this is one of the most likely applications of Fowler-Nordheim plots.]

It subsequently became clear that the original thinking above is strictly correct only for the physically unrealistic situation of a flat emitter and an exact triangular barrier. For real emitters and real barriers a "slope correction factor" σ_{FN} has to be introduced, yielding the revised formula

$$S_{FN} = -\sigma_{FN} b\phi^{3/2}/\beta \tag{49}$$

The value of σ_{FN} will, in principle, be influenced by any parameter in the physically complete Fowler-Nordheim-type equation for $i(V)$ that has a voltage dependence.

At present, the only parameter that is considered important is the correction factor v_F relating to the barrier shape, and the only barrier for which there is any well-established detailed theory is the Schottky-Nordheim barrier. In this case, σ_{FN} is given by a mathematical function called s. This function s was first tabulated correctly (as a function of the Nordheim parameter y) by Burgess, Kroemer and Houston in 1953; and a modern treatment that gives s as function of the scaled barrier field f for a Schottky-Nordheim barrier is given in. However, it has long been clear that, for practical emitter operation, the value of s lies in the range 0.9 to 1.

In practice, due to the extra complexity involved in taking the slope correction factor into detailed account, many authors (in effect) put $\sigma_{FN} = 1$

in eq. (49), thereby generating a systematic error in their estimated values of β and/or γ, thought usually to be around 5%.

However, empirical equation (42) – which in principle is more general than Fowler-Nordheim-type equations - brings with it possible new ways of analyzing field emission *i-V* data. In general, it may be assumed that the parameter *B* in the empirical equation is related to the unreduced height *H* of some characteristic barrier seen by tunneling electrons by

$$B = bH^{3/2} / \beta \tag{50}$$

(In most cases, but not necessarily all, *H* would be equal to the local work-function; certainly this is true for metals.) The issue is how to determine the value of *B* by experiment. There are two obvious ways. (1) Suppose that eq. (43) can be used to determine a reasonably accurate experimental value of κ, from the slope of a plot of form [–dln{*i*}/d(1/*V*) vs. *V*]. In this case, a second plot, of $\ln(i)/V^{\kappa}$ vs. 1/*V*, should be an exact straight line of slope –*B*. This approach should be the most accurate way of determining *B*.

(2) Alternatively, if the value of κ is not exactly known, and cannot be accurately measured, but can be estimated or guessed, then a value for *B* can be derived from a plot of the form [ln{*i*} vs. 1/*V*]. This is the form of plot used by Millikan and Lauritsen in 1928. Re-arranging eq. (43) gives

$$B = -\,\mathrm{d}\ln(i) / \mathrm{d}(1/V) - \kappa(1/V) \tag{51}$$

Thus, *B* can be determined, to a good degree of approximation, by determining the mean slope of a Millikan-Lauritsen plot over some range of values of 1/*V*, and by applying a correction, using the value of 1/*V* at the midpoint of the range and an assumed value of κ.

The main advantages of using a Millikan-Lauritsen plot, and this form of correction procedure, rather than a Fowler-Nordheim plot and a slope correction factor, are seen to be the following. (1) The plotting procedure is marginally more straightforward. (2) The correction involves a physical parameter (*V*) that is a measured quantity, rather than a physical parameter (*f*) that has to be calculated [in order to then calculate a value of *s*(*f*) or, more generally $\sigma_{FN}(f)$]. (3) Both the parameter κ itself, and the correction procedure, are more transparent (and more readily understood) than the Fowler-Nordheim-plot equivalents. (4) This procedure takes into account all physical effects that influence the value of κ, whereas the Fowler-Nordheim-plot correction procedure (in the form in which it has been carried out for the last 50 years) takes into account only those effects associated with barrier shape – assuming, furthermore, that this shape is that of a Schottky-Nordheim barrier. (5) There is a cleaner separation of theoretical and technological concerns: theoreticians will be interested in establishing what information any measured values of κ provide about CFE theory; but experimentalists can simply use measured values of κ to make more accurate estimates (if needed) of field enhancement factors. This correction procedure for Millikan-Lauritsen plots will become easier to apply when a sufficient number of measurements of κ have been made, and a better idea is available of what typical values actually are. At present, it seems probable that for most materials κ will lie in the range $-1<\kappa<3$.

Further Theoretical Information

Developing the approximate theory of CFE from metals above is comparatively easy, for the following reasons.

(1) Sommerfeld's free-electron theory, with its particular assumptions about the distribution of internal electron states in energy, applies adequately to many metals as a first approximation.

(2) Most of the time, metals have no surface states and (in many cases) metal wave-functions have no significant "surface resonances".

(3) Metals have a high density of states at the Fermi level, so the charge that generates/screens external electric fields lies mainly on the outside of the top atomic layer, and no meaningful "field penetration" occurs.

(4) Metals have high electrical conductivity: no significant voltage drops occur inside metal emitters: this means that there are no factors obstructing the supply of electrons to the emitting surface, and that the electrons in this region can be both in effective local thermodynamic equilibrium and in effective thermodynamic equilibrium with the electrons in the metal support structure on which the emitter is mounted.

(5) Atomic-level effects are disregarded.

The development of "simple" theories of field electron emission, and in particular the development of Fowler-Nordheim-type equations, relies on all five of the above factors being true. For materials other than metals (and for atomically sharp metal emitters) one or more of the above factors will be untrue.

For example, crystalline semiconductors do not have a free-electron-like band-structure, do have surface states, are subject to field penetration and band bending, and may exhibit both internal voltage drops and statistical decoupling of the surface-state electron distribution from the electron distribution in the surface region of the bulk band-structure (this decoupling is known as "the Modinos effect").

This, basically, is why this article is confined to the theory of CFE from bulk metals. The complications involved in presenting CFE theory for non-metals are too great to be dealt with satisfactorily via Wikipedia. In any case, the basic theory of CFE from bulk metals needs to be understood first.

In practice, the theory of the actual Fowler-Nordheim tunneling process is much the same for all materials (though details of barrier shape may vary, and modified theory has to be developed for initial states that are localized rather than are travelling-wave-like). However, notwithstanding such differences, one expects (for thermodynamic equilibrium situations) that all CFE equations will have exponents that behave in a generally similar manner. This is why applying Fowler-Nordheim-type equations to materials outside the scope of the derivations given here often works. If interest is only in parameters (such as field enhancement factor) that relate to the slope of Fowler-Nordheim or Millikan-Lauritsen plots and to the exponent of the CFE equation, then Fowler-Nordheim-type theory will often give sensible estimates. However, attempts to derive meaningful current density values will usually or always fail.

Note that a straight line in a Fowler-Nordheim or Millikan-Lauritsen plot does *not* indicate that emission from the corresponding material obeys a Fowler-Nordheim-type equation: it indicates only that the emission mechanism for individual electrons is probably Fowler-Nordheim tunneling.

Different materials may have radically different distributions in energy of their internal electron states, so the process of integrating current-density contributions over the internal electron states may give rise to significantly different expressions for the current-density pre-exponentials, for different classes of material. In particular, the power of barrier field appearing in the pre-exponential may be different from the original Fowler-Nordheim value "2". Investigation of effects of this kind is an active research topic. Atomic-level "resonance" and "scattering" effects, if they occur, will also modify the theory.

Where materials are subject to field penetration and band bending, a necessary preliminary is to have good theories of such effects (for each different class of material) before detailed theories of CFE can be developed. Where voltage-drop effects occur, then the theory of the emission current may, to a greater or lesser extent, become theory that involves internal transport effects, and may become very complex.

3

Thin Film Solar Cell

A thin-film solar cell (TFSC), also called a thin-film photovoltaic cell (TFPV), is a solar cell that is made by depositing one or more thin layers (thin film) of photovoltaic material on a substrate. The thickness range of such a layer is wide and varies from a few nanometres to tens of micrometres.

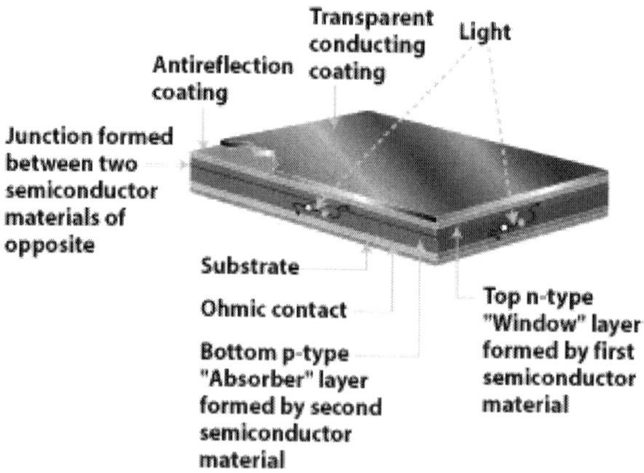

Figure: *Cross-section of a thin-film polycrystalline solar cell.*

Many different photovoltaic materials are deposited with various deposition methods on a variety of substrates. Thin-film solar cells are usually categorized according to the photovoltaic material used:

- Amorphous silicon (a-Si) and other thin-film silicon (TF-Si)
- Cadmium telluride (CdTe)
- Copper indium gallium selenide (CIS or CIGS)

- Dye-sensitized solar cell (DSC) and other organic solar cells

$$\begin{pmatrix} Cu \\ Ag \\ Au \end{pmatrix}\begin{pmatrix} Al \\ Ga \\ In \end{pmatrix}\begin{pmatrix} S \\ Se \\ Te \end{pmatrix}_2$$

Possible combinations of Group-(XI, XIII, XVI)
elements in the periodic table that yield
a compound showing photovoltaic effect
(*Cu, Ag, Au* | *Al, Ga, In* | *S, Se, Te*).

Initially appearing as small strips powering hand-held calculators, thin-film PV is now available in very large modules used in sophisticated building-integrated installations and vehicle charging systems. GBI Research, a private company selling reports it calls "global business intelligence", projects thin film production to grow 24% from 2009 levels and to reach 22,214 MW in 2020. "Expectations are that in the long-term, thin-film solar PV technology would surpass dominating conventional solar PV technology, thus enabling the long sought-after grid parity objective."

Thin-Film Silicon

A silicon thin-film cell uses amorphous (a-Si or a-Si:H), protocrystalline, nanocrystalline (nc-Si or nc-Si:H) or black silicon. Thin-film silicon is opposed to *wafer* (or *bulk*) silicon (monocrystalline or polycrystalline).

Design and Fabrication

The silicon is mainly deposited by chemical vapor deposition, typically plasma-enhanced (PE-CVD), from silane gas and hydrogen gas. Other deposition techniques being investigated include sputtering and hot wire techniques.

The silicon is deposited on glass, plastic or metal which has been coated with a layer of transparent conducting oxide (TCO).

A p-i-n structure is usually used, as opposed to an n-i-p structure. This is because the mobility of electrons in a-Si:H is roughly 1 or 2 orders of magnitude larger than that of holes, and thus the collection rate of electrons moving from the n- to p-type contact is better than holes moving from p- to n-type contact. Therefore, the p-type layer should be placed at the top where the light intensity is stronger, so that the majority of the charge carriers crossing the junction would be electrons.

Micromorphous Silicon

Micromorphous silicon module technology combines two different types of silicon, amorphous and microcrystalline, in a top and a bottom photovoltaic cell. These two materials are chosen because their different absorption spectrums and easily combined process. Because the two different materials are both Si, they can be manufactured in the same technology, which now is PECVD. The band gap of a-Si is 1.7 eV and that of c-Si is 1.1 eV, which eventually broaden the spectral acceptance of the micromorph tandem solar cell.

The c-Si layer can help to absorb the energy of red and infrared spectrum and increase the overall efficiency. The best efficiency can be achieved at transition between a-Si and c-Si. Use of protocrystalline silicon for the intrinsic layer has shown to optimize the open-circuit voltage of an a-Si photovoltaic cell.

Efficiency

These types of silicon present dangling and twisted bonds, which results in deep defects (energy levels in the bandgap) as well as deformation of the valence and conduction bands (band tails). The solar cells made from these materials tend to have lower energy conversion efficiency than bulk silicon (also called crystalline or wafer silicon), but are also less expensive to produce. The quantum efficiency of thin-film solar cells is also lower due to reduced number of collected charge carriers per incident photon.

Amorphous silicon has a higher bandgap (1.7 eV) than crystalline silicon (c-Si, 1.1 eV), which means it absorbs the visible part of the solar spectrum more strongly than the infrared portion of the spectrum. As nc-Si has about the same bandgap as c-Si, the nc-Si and a-Si can advantageously be combined in thin layers, creating a layered cell called a *tandem cell*. The top cell in a-Si absorbs the visible light and leaves the infrared part of the spectrum for the bottom cell in nc-Si.

Recently, solutions to overcome the limitations of thin-film silicon have been developed. Light trapping schemes where the incoming light is obliquely coupled into the silicon and the light traverses the film several times enhance the absorption of sunlight in the films. Thermal processing techniques enhance the crystallinity of the silicon and pacify electronic defects.

Building Integrated Photovoltaics

Thin film solar panels are commercially available for installation onto the roofs of buildings, either applied onto the finished roof, or integrated into the roof covering. The advantage over traditional PV panels is that they are very low in weight, are not subject to wind lifting, and can be walked on (with care). The comparable disadvantages are increased cost and reduced efficiency.

A silicon thin film technology is being developed for building integrated photovoltaics (BIPV) in the form of semitransparent solar cells which can be applied as window glazing. These cells function as window tinting while generating electricity.

Figure: *Thin film photovoltaic laminates being installed onto a roof.*

Organic Solar Cells

An organic solar cell would use a photovoltaic polymer compound to produce electricity. Current research devices show low efficiency and stability. If an organic solar cell of usable efficiency can be developed, it may reduce the cost of PV installations.

Efficiencies, Volumes and Prices

Since the invention of the first modern silicon solar cell in 1954, incremental improvements have resulted in modules capable of converting 12 to 18 percent of solar radiation into electricity.

The performance and potential of thin-film materials are high, reaching cell efficiencies of 12–20%; prototype module efficiencies of 7–13%; and production modules in the range of 9%. Future module efficiencies are expected to climb close to the state-of-the-art of today's best cells, or to about 10–16%.

Annual manufacturing volume in the United States has grown from about 12 megawatts (MW) per year in 2003 to more than 20 MW/yr in 2004; 40–50 MW/yr production levels were expected in 2005 with continued rapid growth in the years after that.

Costs are expected to drop to below \$100/m^2 in volume production, and could reach even lower levels—well under \$50/m^2, the DOE/NREL goal for thin films—when fully optimized.

At these levels, thin-film modules will cost less than fifty cents per watt to manufacture, opening new markets such as cost-effective distributed power and utility production to thin-film electricity generation.

As crystalline silicon price rose, the production cost of crystalline silicon-based solar cell module in 2008 was at some point 4–5 times higher than that of thin film modules. Thin-film producers still enjoy in 2009 price advantage as its production cost is 20% less than that of silicon modules. It is expected that the production cost of thin-film will continue dropping (40% less than silicon), as Chinese producers are now putting more resources into R&D and partnering with manufacturing equipment suppliers.

Production, Cost and Market

In recent years, the manufacturers of thin-film solar modules have brought costs down and gained competitive strength through advanced thin film technology. However, the traditional crystalline silicon technologies will not give up their market positions for a few years because they still hold considerable development potential in terms of the cost. Efficiency of thin film solar is considerably lower and thin film solar manufacturing equipment suppliers intend to score costs of below USD 1/W, and Anwell Technologies Limited claimed that they intend to bring it down further to USD 0.5/W. In July 2013 the price of Thin film a-Si dropped to €0.38/Wp, about USD 0.502493. Those equipment suppliers have been doing R&D for micro-morphous silicon modules since 2008. This technology represents a development based on the thin-film panels made of ordinary amorphous silicon marketed at present that brings higher cell efficiency by depositing an additional absorber layer made of micro crystalline silicon on the amorphous layer. Some equipment suppliers even claim that there will be machinery in market to manufacture these new modules at $0.70. With such potential of further development of thin film solar technology, the European Photovoltaic Industry Association (EPIA) expects that manufacturing capacities for these technologies will double to over 4GW by 2010 representing a market share of around 20%. In 2011, GE announced plans to spend $600 million on a new CdTe solar cell plant and enter this market.

Potential thin film manufacturers faced severe price competition from Chinese refiners of silicon and manufacturers of conventional solar panels. As of January, 2013 one of the consequences was that some firms developing thin film technology together with their patents have been sold to Chinese firms below cost.

Installations

First Solar, the CdTe thin-film manufacturer stated that "at the end of 2007, over 300 MW of First Solar PV modules had been installed worldwide." Below is a list of several recent installations:

- Since 16 October 2008, Germany's largest thin-film pitched roof system, constructed by Riedel Recycling, has been in operation and producing solar power in Moers near Duisburg. Over eleven thousand cadmium telluride modules, from First Solar, deliver a total of 837 kW.

- First Solar recently completed a 2.4 MW rooftop installation as part of Southern California Edison program to install 250 MW of rooftop solar panels throughout Southern California over by 2013.
- First Solar announced a 7.5 MW system to be installed in Blythe, CA, where the California Public Utilities Commission has accepted a 12 ¢/kWh power purchase agreement with First Solar (after the application of all incentives).
- Construction of a 10 MW plant in the Nevada desert began in July 2008. First Solar is partnering with Sempra Generation, which will own and operate the PV power-plant, being built next to their natural gas plant.
- Stadtwerke Trier (SWT) in Trier, Germany is expected to produce over 9 GWh annually
- A 40 MW system is being installed by Juwi in Waldpolenz Solar Park, Germany. At the time of its announcement, it was both the largest planned and lowest cost PV system in the world. The price of 3.25 euros translated then (when the euro was equal to US$1.3) to $4.2 per installed watt.
- 4.8KW of thin film flexible solar panels manufactured by Uni-Solar Ovonic installed on a South Beach hurricane-prone residence in 2008.

Denver-based Conergy Americas and officials at California's South San Joaquin Irrigation District (SSJID) have installed what is believed to be the world's first single-axis solar tracking system featuring thin-film photovoltaic cells.

Flexible Solar Cells Made of Nanowires Microwires

Current solar cell technologies are dominated by silicon. Limited primarily by silicon's inherent properties, these solar cells convert only 15–20% of solar energy into electricity. Solar cells made from III-V compound semiconductors (materials that contain group III and group V elements) have much higher efficiencies, due to their better optical (absorption) and electrical (charge mobility) properties. Furthermore, with different III-V materials on the same device, researchers can produce multiple junctions or heterostructures to cover a broader range of the solar spectrum. In these devices, researchers have demonstrated efficiencies of over 40% under concentrated solar energy.

The additional benefits afforded by nanotechnology may further enhance solar cell performance. Nanowire research has emerged as a quickly growing field. Much excitement stems from the unique electronic and optical properties of semiconductor nanowires. Nanowires of a particular substance may behave quite unlike much larger 'bulk' samples of the same material. Nanowires are of great interest in photovoltaics because of their large surface area, high aspect ratio (long and thin), intrinsic antireflection effect (which increases light absorption), and ability to direct light absorption with specifically designed arrays.

More importantly, core-shell nanowires, which operate based on radial p-n junctions, represent a revolution in photovoltaics by decoupling light absorption from carrier collection pathways: light is absorbed vertically, whereas carriers are separated radially. This separation eliminates the once-fundamental trade-off between light absorption and carrier collection.

By incorporating the superior photovoltaic properties of III-V semiconductors into nanowire structures, researchers expect to achieve efficiencies similar to today's best solar cells, with significantly less material. Because nanowire-based cells use less material than planar devices, such changes can ultimately reduce costs.

Figure: *Scanning electron microscopy (SEM) image of gallium arsenide/ aluminum gallium arsenide (GaAs/AlGaAs) core-shell nanowires.*

Figure depicted above, shows a scanning electron microscopy image of typical gallium arsenide/aluminum gallium arsenide (GaAs/ AlGaAs) core-shell nanowires used to fabricate solar cells. These nanowires are synthesized on GaAs platforms using the vapor-liquid-solid process—a technique that relies on gold nanoparticles to catalyze nanowire growth—in a metal organic chemical vapor deposition system.

By optimizing growth conditions, we have significantly improved the crystal quality of the GaAs nanowires. Furthermore, by growing a high-quality AlGaAs shell surrounding the GaAs nanowire, we achieved excellent optical quality and long carrier lifetimes.

These properties are very desirable for solar cells because carriers can otherwise recombine easily at the surface and reduce the output current.

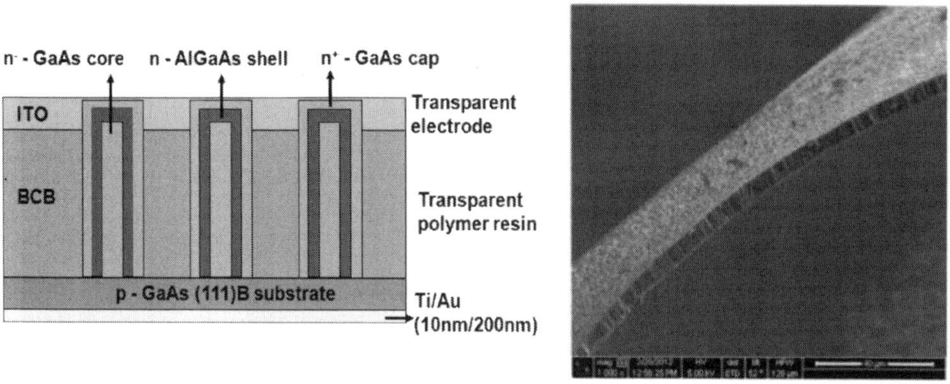

***Figure:** Left: Schematic diagram of a nanowire solar cell. Right: SEM image of BCB planarized nanowires peeled off from the substrate. Ti: Titanium. Au: Gold. BCB: Benzocyclobutene. ITO: Indium tin oxide. p, n, n^+, n^-: Different types of semiconductor materials. (111) refers to the crystal orientation of the substrate.*

Since nanowires are non-planar structures, fabrication of nanowire devices requires non-standard semiconductor manufacturing processes. Figure shows a schematic diagram of a prototype core-shell GaAs/AlGaAs nanowire solar cell and the necessary fabrication processes.

The p-type semiconductor GaAs substrate and the n-type GaAs nanowires form a substrate-nanowire p-n junction device. Benzocyclobutene (BCB) resist, applied via spin coating and thermal curing, planarizes the vertically standing nanowires on the substrate.

We chose BCB for its excellent planarizing characteristics, such as uniform coverage, and its excellent transmission and insulating properties. Inductively coupled plasma reactive ion etching with a sulfur hexafluoride and oxygen (SF_6/O_2) gas mixture then etches back the BCB to expose the nanowire tips for top contacts. Finally, sputtering and electron beam evaporation form a 200nm-thick transparent indium tin oxide layer and a titanium/gold (10nm/200nm) layer, as top n- and bottom p-contacts, respectively.

This technique holds great promise for creating flexible and lightweight nanowire solar cells. Reduced material usage and substrate re-use should also help to lower costs. Such solar cells could be integrated into, rather than installed on, surfaces such as clothes and fabrics.

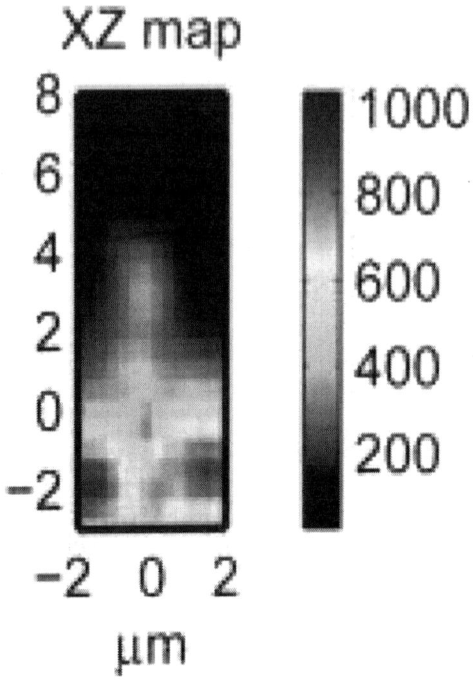

Figure: *Photocurrent map along the growth direction of a single GaAs/AlGaAs nanowire, using the TOBIC (two-photon induced current) technique. Photocurrent generated from both a nanowire and the substrate below it can be clearly identified for in-depth study of carrier generation and collection.*

To understand what is happening within the nanowire solar cells, we developed a novel technique based on two-photon induced current. The technique provides a 3D map of the photocurrent from each nanowire at submicrometre resolution. Using 'below bandgap' (two-photon) laser excitation, carriers can be generated specifically at the focal voxel and not above or below this point within the focus cone (as would occur for above-gap excitation).

Figure shows photocurrent mapping along the growth direction of a single nanowire from our solar cell, revealing efficiency hotspots, carrier collection pathways, and recombination mechanisms with high spatial resolution.

This technique opens up enormous opportunity to explore various nanowire solar cell designs and further exploit the unique properties of 1D nanostructures in high-efficiency photovoltaic devices. Nanowire solar cells show great promise for next-generation photovoltaics.

However, addressing the many new material and device challenges that arise from these unconventional nanostructures will require more mature simulation, manufacturing, and characterization tools. Researchers must focus their attention on these goals to achieve high-performance devices that will find application in daily life.

Plastic Solar Panel Technology

In the last two decades, mankind has witnessed intensive research in solar panel technology. Various companies, in conjunction with the leading universities and innovation labs of the world have been making attempt to cause an improvement in the solar panel technology so that they could make their products marketable. In this article, we are going to have a look at the Plastic Solar Panel Technology that is developed by a company called Solarmer Energy Inc.

Solarmer Energy Inc. is regarded as one of the innovators in the industry of solar panels. They have caused the development of the transparent, flexible solar panels that are based upon plastic and its derivatives. Thus this company has made a leap in generating renewable energy from the power of the sun.

Solar panels, mainly due to the construction process and also the materials used in it, have a limited number of applications. However, the scenario is slightly altered by Solarmer Energy Inc. They have opened doors to whole lot of applications that is creeping in to use the solar panels that have been developed out of plastic. In the sector of harnessing renewable form of energy, silicon as an element failed to make a mark in the wide range of applications of solar panels. But in this case, the rest of the things are also addressed.

Because now the manufacturing process is catered by the plastic as a raw material, these kinds of solar panels have the potential to reduce the cost at which solar panels used to be available in the market. Statistics speak for themselves. The cost of solar panel with this technology is capable of making the provision of solar panels at 12-15 cents per kWh and typically less than a dollar per Watt of power. All this makes us conclude that this technology of solar panels is par than any other that is momentarily available in the market.

Mainly, the solar panels that are developed using Solarmer's plastic technology address costing issues in the following ways:

Cost of raw material: Low. The grade of plastic is irrespective of the solar power throughput produced by the solar panels. This much more cost efficient than extraction of silicon and the associated processes.

Material usage: Low. Plastics have extraordinary capability to absorb light energy. Hence we need to use only a thin film of plastic per unit production of a solar cell. Typically, only few micrometres thickness is more than sufficient.

Manufacturing cost: Lower than silicon. Thin film plastic is to be printed on the solar cell substrate. This can be compared to printing cost of a newspaper which is too low. Also, the impact on the environment is minimal due to the low temperature fabrication technique used here. This means energy consumption is also low, this further reducing the overall solar panel cost in the market.

Organic plastic is used for the purpose of this variant of solar cell. In terms of the dimensions too, the plastic solar panels are compact, light in weight and also find applications where flexibility is of utmost importance. Traditional solar cell just fails to address applications where flexibility requirement issue is present. Since a wide range of colours can be used in the plastic manufacturing process, solar cells are available in a variety of colours.

Currently, solar panels using plastic technology mentioned above are entering the market. The technology behind it promises that the entry in the market and the rate at which the user acceptance will be there is something to look out for. In the future, plastic solar panels will promise to be of same or even better efficiency, and wipe out the market share of silicon solar panel industry that currently holds the market capitalization of 90%. This will also ease the integration of solar panels with the other devices that make up a photovoltaic system.

Wet Chemical Growth and Applications in Energy Conversion

Wet chemistry is a term used to refer to chemistry generally done in the liquid phase. It is also known as bench chemistry because many of the tests performed are done at a lab bench.

Traditionally, it involves the use of laboratory glassware, such as beakers and flasks, and excludes quantitative chemical analysis using instrumentation. Many high school and college laboratories teach students basic wet chemistry methods. Before the age of theoretical and computational chemistry it was the predominant form of scientific discovery in the chemical field. This is why it is sometimes referred to as classic chemistry or classical chemistry. Because of the high volume of wet chemistry that must be done in today's society and quality control requirements, many wet chemistry methods have been automated and computerized for streamlined analysis.

Wet chemistry techniques can be used for qualitative chemical measurements, such as changes in colour (colorimetry), but often involves more quantitative chemical measurements, using methods such as gravimetry and titrimetry.

Energy Transformation

Energy transformation or energy conversion is the process of changing one form of energy to another. In physics, the term energy describes the capacity to produce certain changes within a system, without regard to limitations in transformation imposed by Entropy. Changes in total energy of systems can only be accomplished by adding or removing energy from them, as energy is a quantity which is conserved (unchanging), as stated by the first law of thermodynamics. Mass-energy equivalence, which arose from special relativity, says that changes in the energy of systems will also coincide with changes (often small in practice) in the system's mass, and the mass of a system is a measure of its energy content.

Energy in its most various forms may be used in natural processes, or to provide some service to society such as heating, refrigeration, light, or performing mechanical work to operate machines. For example, an internal combustion engine converts the potential chemical energy in gasoline and oxygen into thermal energy which, by causing pressure and performing work on the pistons, is transformed into the mechanical energy that accelerates the vehicle (increasing its kinetic energy). A solar cell converts the radiant energy of sunlight into electrical energy that can then be used to light a bulb or power a computer.

The generic name for a device which converts energy from one form to another, is a *transducer*.

Conversions to thermal energy (thus raising the temperature) from other forms of energy, may occur with essentially 100% efficiency (many types of friction do this). Conversion among non-thermal forms of energy may occur with fairly high efficiency, though there is always some energy dissipated thermally due to friction and similar processes. Sometimes the efficiency is close to 100%, such as when potential energy is converted to kinetic energy as an object falls in vacuum, or when an object orbits nearer or farther from another object, in space.

On the other hand, conversion of thermal energy to other forms, thus reducing the temperature of a system, has strict limitations, often keeping its efficiency much less than 100% (even when energy is not allowed to escape from the system). This is because thermal energy has already been partly spread out among many available states of a collection of microscopic particles constituting the system, which can have enormous numbers of possible combinations of momentum and position (these combinations are said to form a phase space). In such circumstances, a measure called entropy, or evening-out of energy distributions, dictates that future states of an isolated system must be of at least equal evenness in energy distribution. In other words, there is no way to *concentrate* energy without spreading out energy somewhere else.

Thermal energy in equilibrium at a given temperature already represents the maximal evening-out of energy between all possible states. Such energy is sometimes considered "degraded energy," because it is not entirely convertible a "useful" form, i.e. one that can do more than just affect temperature. The second law of thermodynamics is a way of stating that, for this reason, thermal energy in a system may be converted to other kinds of energy with efficiencies approaching 100%, only if the entropy (even-ness or disorder) of the universe is increased by other means, to compensate for the decrease in entropy associated with the disappearance of the thermal energy and its entropy content. Otherwise, only a part of thermal energy may be converted to other kinds of energy (and thus, useful work), since the remainder of the heat must be reserved to be transferred to a thermal reservoir at a lower temperature, in such a way that the increase in Entropy for this process more than compensates for the entropy decrease associated with transformation of the rest of the heat into other types of energy.

Flexible Energy Storage Devices Using Nanomaterials

Recently, great interest has been aroused in flexible/bendable electronic equipment such as rollup displays and wearable devices. As flexible energy conversion and energy storage units with high energy and power density represent indispensable components of flexible electronics, they should be carefully considered. However, it is a great challenge to fabricate flexible/ bendable power sources. This is mainly due to the lack of reliable materials that combine both electronically superior conductivity and mechanical flexibility, which also possess high stability in electrochemical environments. In this work, we report a new approach to flexible energy devices. We suggest the use of a flexible electrode based on free-standing graphene paper, to be applied in lithium rechargeable batteries.

Moreover graphene paper is a functional material, which does not only act as a conducting agent, but also as a current collector. The unique combination of its outstanding properties such as high mechanical strength, large surface area, and superior electrical conductivity make graphene paper, a promising base material for flexible energy storage devices. In essence, we discover that the graphene based flexible electrode exhibits significantly improved performances in electrochemical properties, such as in energy density and power density. Moreover graphene paper has better life cycle compared to non-flexible conventional electrode architecture. Accordingly, we believe that our findings will contribute to the full realisation of flexible lithium rechargeable batteries used in bendable electronic equipments.

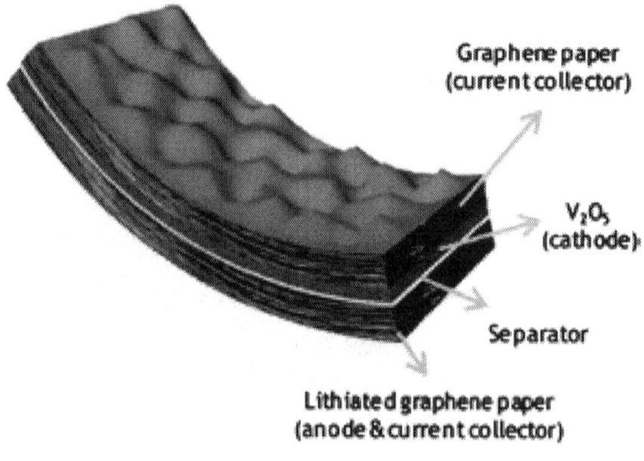

Paper-based Energy Storage Devices

Paper based batteries and supercapacitors comprising composites composed of cellulose from the green algae Cladophora and the electronically conducting polymer polypyrrole are developed in a collaboration between the department of Materials Chemistry and the Department of Nanotechnology and Functional Materials at Uppsala University. The devices, which are light-weight and inexpensive, contain two electrodes made of cellulose sheets coated with conducting polymer, separated by an insulating cellulose sheet (soaked with a sodium chloride solution), as the separator. Due to the large surface area of the Cladophora cellulose (i.e. 100 m^2/g) and the conformal coating of the cellulose fibres with a 50 nm thin layer of polypyrrole, the devices have a relatively high capacity as well as power capability. The charging time can be made as short as about 10 s. These properties and the absence of metals render these devices interesting for many new applications as these batteries and supercapacitors constitute promising complements to contemporary energy storage devices. Research is currently carried out to improve the charge storage properties of the composite and to identify and develop applications based on this novel technology.

Rechargeable Lithium-ion (Li-Ion) Battery Technology

Rechargeable Lithium-ion (Li-Ion) battery technology has come a long way since its introduction in the early 1990s. Over the last two decades, they have become the technology of choice for powering portable electronic devices such as cellular phones and laptop computers. Currently, Lithium-ion batteries are steadily replacing Nickel-Cadmium (NiCd) and Nickel-Metal Hydride (NiMH) battery technologies in portable power tools. In the future, Lithium-ion batteries are poised to power a new generation of hybrid electric vehicles (HEV), plug-in hybrids (PHEV) and electric vehicles (EV). Another emerging application for Lithium-ion technology is in battery electrical energy storage systems for smart grids that are powered by traditional energy sources like coal, as well as intermittent renewable energy sources like solar and wind.

The optimum combination of long run time (high energy density at the desired power) and long life (recharge characteristics) sets Lithium-ion battery technology apart from the competition. Needless to say, safety and cost expectations of the application also have to be satisfied. While all of these requirements are being met for portable electronics, the technology is just beginning to move up the optimization curve for emerging applications such as electric automotives, power tools and storage systems. For example, one factor that distinguishes the portable electronics and the electric vehicle application is power density. The latter requires much higher charge and discharge rates compared to the former. While higher power can be achieved to some extent by redesigning the way the battery cell is constructed, nanomaterials are also expected to play a key role in the achievement of high-power capability. Nanomaterial strategies are also being employed to provide better strain accommodation in high-capacity electrodes so that this storage capacity can be extracted reversibly with minimum compromise in cycle life. Alternatively, the stable cycle life and storage life characteristics of "zero-strain" electrode materials can be improved even further by utilising nanosized versions of these electrodes, thereby providing a new generation of electrical energy storage options for smart grids and back-up power systems.

Overview of Lithium-ion Electrode Materials Chemistries

Conventional Lithium-ion battery materials typically start as 10-50 micron sized particles, which are then coated onto aluminum or copper current collectors along with conductivity enhancers and binders. The work-horse cathode chemistry for the last couple of decades has been Lithium Cobalt Oxide, $LiCoO_2$, (Aldrich Prod. No., 442704 a layered compound with a distorted rock-salt (α-$NaFeO_2$) structure. Some alternative cathodes, like the three dimensional spinel Lithium Manganese Oxide, $LiMn_2O_4$, (Aldrich Prod. 725129) have been commercialized only in niche applications, because of performance limitations. Carbon-based materials have been the preferred choice for anodes, with some version of graphite being utilised in a majority of the commercially available batteries. Battery developers choose electrode materials with the intent to optimize performance from the standpoints of energy, power, cycle life, cost and thermal stability. Until recently, much of this optimization has taken place with the portable electronic application in mind. More recently however, a significant effort is also being spent on new and emerging applications like power tools, electric vehicles and battery electrical energy storage systems. New chemistries have emerged for cathodes including a variety of mixed metal oxides, like $LiMn_{1.5}Ni_{0.5}O_4$ (spinel, Aldrich Prod. No. 725110), $LiNi_{0.33}Mn_{0.33}Co_{0.34}O_2$ (layered) and $LiNiCoAlO_2$ (layered) and metal phosphates, like $LiFePO_4$, $LiCoPO_4$ (Aldrich Prod. No. 725145) and $LiMnPO_4$, (olivine). For anodes, new materials include oxides such as Lithium Titanates ($Li_4Ti_5O_{12}$) & Tin Oxide (SnO_2, Aldrich Prod. No. 549657), elemental silicon (Si, Aldrich Prod. No. 267414) & tin (Sn, Aldrich Prod. No. 265640) and many carbon-based materials. In many of these new chemistries, having the materials in nanoparticle form or as a nanostructured particle or film is critical to achieving the desired performance dictated by the end application.

Advantages of Nanomaterials for Lithium-ion Applications

From the battery application perspective, the incentive for implementing a nanomaterial electrode as a Lithium-ion storage material would be to derive significant improvement in energy, power, cycle life or some combination of the same. Nanoparticles or nanopowder electrode materials, i.e., ultrafine versions of the conventional micron-sized electrode powders, are the earliest implementation of nanomaterials science in the Lithium-ion battery application. Indeed, carbon-black, a nanomaterial that has been around for several decades, has been used in Lithium-ion batteries since its early days. While carbon-black is used in the electrode, it does not store electrical energy and merely acts as a "passive" conductivity enhancer to improve power capability. However, by designing the "active" energy storage component of the electrode as a nanoparticle, significant performance improvements can be realised for two reasons:

1. shorter diffusion lengths for the Lithium-ion travelling from the particle core to the surface where it transfers to the electrolyte and
2. higher electrode-electrolyte contact area arising from the inherently high surface areas characteristic of the particles.

Reducing electrode particle size into the nanoscale regime is also believed to substantially reduce the mechanical stresses caused by volumetric expansion and contraction during charge and discharge. A recently developed model suggests that particles must have a size under a certain critical radius so that the strain produced by intercalation can be accommodated elastically, rather than by plastic deformation, allowing for the full recovery of the original, stable structure. Examples of nanomaterials either in nanopowder form or as nanostructured films with wire, rod, whisker or columnar morphologies are actively being pursued to maximize cycle life with minimum compromise in energy density.

Some processing advantages may also be realised while working with nanosized Li-ion storage electrode materials. For example, most Lithium-ion cathode materials are made from precursors containing lithium and other transition metals, which are combined and then heat treated under a variety of conditions to arrive at the desired oxide or phosphate composition. Heat treatment processes can be tedious and energy-intensive, particularly for large particle aggregates where the surface of the aggregate may experience a different thermal profile compared to the bulk. Improper heat treatment leads to nonhomogeneous composition in the material and hence deterioration in performance. While heat treating nanomaterials, it is easier to maintain a homogenous thermal profile throughout the material and thereby produce homogenous compositions without having to resort to energy-intensive processing methods.

Nanomaterials could also enable ultrathin and flexible electrode geometries, which may spawn a new generation of high-rate battery formats for low profile components such as sensors, RFID and flexible devices for consumer and medical applications. Dispersions and inks made using nanopowder electrodes could be used to fabricate low-cost printed batteries via roll-to-roll processing or to integrate thin-film batteries with other devices in printed electronic assemblies made via ink-jet processes.

Classification and Production of Nanomaterials

The term nanomaterial is typically used to refer to materials that have at least one dimension of less than 100 nm. In using the term, there is also an underlying implication that the material has some enhanced property or characteristic compared to larger particle size versions of the same composition. Nanomaterials are not a new class of materials, even though the recent attention that they have been garnering may suggest otherwise. Some of these materials, like volcano dust, have existed in nature forever. Other man-made materials, like carbon black and fumed titania (TiO_2), have been around for several decades. One way to classify nanomaterials is to categorize them according to the method by which they are produced, i.e., physical or chemical. The physical method can be further sub-divided into mechanical or phase-change methods. In the physical-mechanical method, particles in the nanometre size range are produced by milling or grinding larger particles of the target composition without any accompanying chemical change.

This method is commonly referred to as a "top-down" approach to making nanomaterials. In physical-phase change methods, the nanomaterial is created via a phase change process. Examples include direct precipitation, in which a material in solution is precipitated as a solid nanomaterial, and thermal, plasma or laser ablation processing in which vaporized material is condensed into solid nanoparticles. In using either of the physical processes, no chemical change is necessary to arrive at the target nanomaterial composition. Chemical methods include processes where the nanomaterial is synthesized from an initial material that is chemically different from the target composition. The target composition is arrived at via chemical synthesis of a solid that is formed directly in the nanoscale. Examples include flame pyrolysis, spray pyrolysis and wet-chemical methods such as sol-gel and solvo-thermal synthesis. The chemical method and the physical-phase change method are both examples of the "bottom-up" approach to making nanomaterials.

An example of a chemical process for making nanomaterials is NanoSpray Combustion, invented by one of the authors, where nanoparticles or nanopowders are produced via combustion chemical vapor condensation (nCCVC). In this process, nanoparticles are produced by the combustion of a solution nanospray that consists of precursors that contain the elements that eventually make up the nanoparticle. nGimat2 s proprietary Nanomiser® device is critical to generating the nanospray from which vapor species are formed in the flame. Condensation of particles can take place either in dry form nanopowders or into a liquid medium to form dispersions. By tailoring precursor formulations and process parameters, nCCVC can be also be used to produce nanomaterials with dopants and surface carbon-coatings to improve performance in the end application.

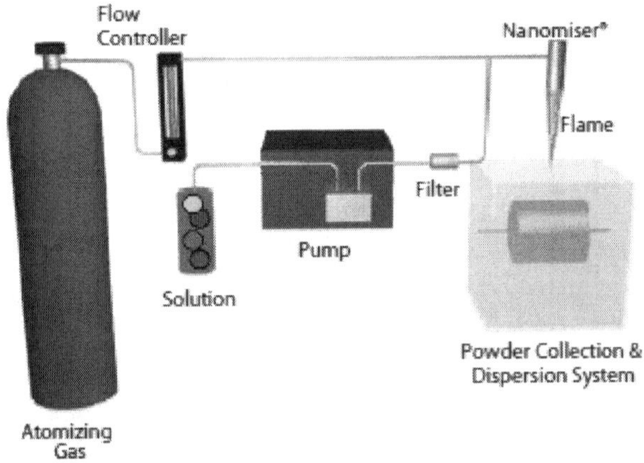

Figure: *Schematic of the NanoSpray combustion chemical vapor condensation (nCCVC) process for making nanopowders.*

While nanoparticles or nanopowders are by definition less than 100 nm in size, other nanomaterials may exhibit structural features that are <100 nm, while existing as larger aggregates like secondary particles or films. For example, lithographic patterning, chemical vapor deposition and physical vapor deposition are all capable of producing nanostructured films with complex structures and are often referred to as form-in-place techniques. A variation of nCCVC, called combustion chemical vapor deposition (nCCVD) can also be classified in this category. This process has been used to make nanostructured thin film coatings with super-hydrophobic, antimicrobial, oxygen-barrier and moisture-barrier characteristics. The nCCVD process has also been used to deposit battery electrode materials. In the following sections, we will focus on the application of nanoparticles or nanopowders in Lithium-ion batteries.

Lithium-ion Battery Application of Nanopowders Made by nCCVC

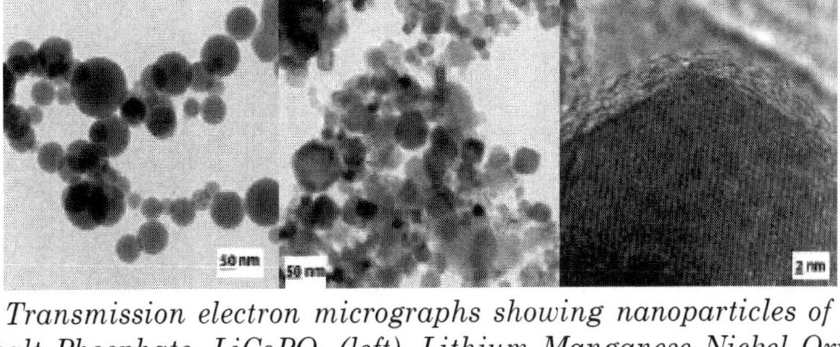

Figure: *Transmission electron micrographs showing nanoparticles of Lithium Cobalt Phosphate, $LiCoPO_4$ (left), Lithium Manganese Nickel Oxide, $LiMn_{1.5}Ni_{0.5}O_4$ (centre), and Lithium Manganese Nickel Oxide, $LiMn_{1.5}Ni_{0.5}O_4$ (right) at higher magnification.*

Using the nCCVC process described above, it is possible to produce a variety of Lithium-ion electrode material compositions that are being pursued by the battery community. The nCCVC process is very wellsuited for making metal oxides and metal phosphate materials. Figure depicted earlier, shows the transmission electron micrographs (TEM) of two representative compounds, $LiCoPO_4$ and $LiMn_{1.5}Ni_{0.5}O_4$. The two materials show different particle morphologies; the former shows spherical features, while the latter is composed of faceted particles. In both cases, the average particle size is well under 100 nm. The higher magnification picture of the $LiMn_{1.5}Ni_{0.5}O_4$ nanoparticle reveals the crystalline structure of the nanoparticle.

Both $LiMn_{1.5}Ni_{0.5}O_4$ and $LiCoPO_4$ are candidates for high-voltage Li-ion cathodes for a new generation of Lithium-ion batteries. For example, $LiMn_{1.5}Ni_{0.5}O_4$ can be charged up to the 4.8–5.0V range compared to 4.2–4.3V charge voltage for $LiCoO_2$ and $LiMn_2O_4$. The higher voltages, combined with the higher theoretical capacity of around 155 mAh/g for LiMn1.5Ni0.5O4 enable a 25% improvement in energy density compared to conventional Lithium-ion cathodes. Other examples of cathode nanomaterials synthesized by nCCVC include $LiCoPO_4$, spinel $LiMn_2O_4$ and orthorhombic $LiMnO_2$ (Aldrich Prod. No. 725137).

The X-ray diffraction (XRD) patterns of the above-mentioned cathode material and representative anode materials made by nCCVC are shown in Figure. All of the patterns match well with reference patterns published by in the literature. Candidate anode nanomaterials made by nCCVC include SnO_2 and $Li_4Ti_5O_{12}$ (Aldrich Prod. No. 702277).

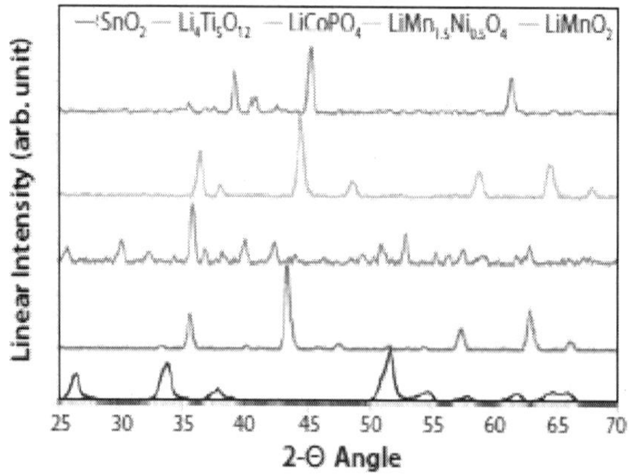

Figure: *X-ray diffraction patterns for Tin Oxide (SnO_2), Lithium Titanate ($Li_4Ti_5O_{12}$), Lithium Cobalt Phosphate ($LiCoPO_4$), Lithium Manganese Nickel Oxide ($LiMn_{1.5}Ni_{0.5}O_4$) and Lithium Manganese Oxide (LiMnO2) nanopowders made with the nCCVC process.*

These materials are targeted for different ends of the applications spectrum. SnO_2 is a candidate for a high energy density anode with theoretical capacities reaching upward of 750 mAh/g and a voltage of <1V vs. lithium metal, compared to 370 mAh/g and <0.5V for the conventional graphite anodes of today. Lower voltages are typically desired for anodes so as to maximize the voltage and energy when coupled with a corresponding higher-voltage cathode. $Li_4Ti_5O_{12}$, on the other hand, has a capacity of around 160 mAh/g at nominal voltages of around 1.5V.17 Although it has a lower energy density, its zero-strain spinel structure can accept many charge-discharge cycles with a minimum loss in capacity. Furthermore, when used in nanopowder form, $Li_4Ti_5O_{12}$ is also known to provide high charge and discharge rate capabilities, as shown below.

Representative discharge voltage profiles for three of the nanopowder electrodes synthesized by nCCVC are shown in Figure below. This data was acquired with small prototype cells which used Lithium-metal as the counter-electrode. The two curves on the upper end of the chart represent 0.1C-rate profiles for cathode materials LiCoPO4 and $LiMn_{1.5}Ni_{0.5}O_4$ and as anticipated both materials demonstrate highvoltage plateaus in the 4.7–4.8V range.

The lower curve represents the C-rate profile for $Li_4Ti_5O_{12}$ and this material demonstrates a flat discharge curve with a plateau at around 1.5V.

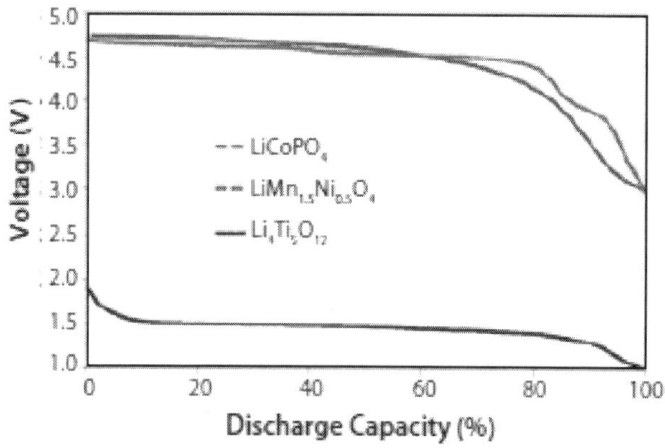

Figure: *Discharge voltage profiles for three representative Lithium-ion storage nanopowders made by nCCVC and assembled in prototype half-cells with a Lithiummetal anode: Lithium Titanate, $Li_4Ti_5O_{12}$ (blue line, bottom) and Lithium Cobalt Phosphate, $LiCoPO_4$ and Lithium Manganese Nickel Oxide, $LiMn_{1.5}Ni_{0.5}O_4$, (red and green lines, top).*

The high-rate discharge capability of the nanomaterials has also been demonstrated in prototype cells for the $Li_4Ti_5O_{12}$ materials. As shown in Figure given earlier, $Li_4Ti_5O_{12}$ can be discharged at 5C and 10C continuous rates and can retain up to 80% and 60% of the C-rate capacities respectively. In comparison, conventional cathode materials like $LiCoO_2$ are typically not recommended for use over 3C-rates. Typical discharge capacities for the $Li_4Ti_5O_{12}$ nanomaterials at C-rate have been around 145 mAh/g. The $Li_4Ti_5O_{12}$ materials have also been cycled for >200 cycles at C-rate. This is highly encouraging performance for the nanomaterial, considering that the test-vehicle is only a lab-assembled prototype cell format. The performance of the material can be expected to improve significantly when assembled using processes that are closer to commercial cell manufacturing methods.

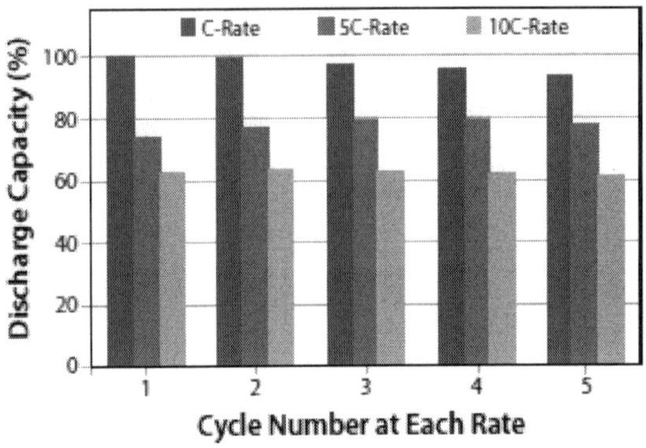

Figure: *Comparison of the relative capacities achieved at high discharge rates (5C and 10C) vs. capacity achieved at 1C discharge rate over 16 cycles for a* $Li_4Ti_5O_{12}$*-Lithiummetal half-cell in prototype format.*

Under the right conditions, the nCCVC process will allow the formation of surface coatings on the electrode particles. Direct coating of the electrode particles with carbon provides higher electronic conductivity to enhance rate capability of the battery. Surface-coated carbon may also eliminate the need for additional conductivity enhancers like carbon-black in the coated electrodes, providing opportunities to increase the energy density of the cell. By tailoring the precursor system and the nCCVC processing conditions, it is possible to make carboncoated $Li_4Ti_5O_{12}$ nanopowders with a single step process. In conventional methods used to process Lithium-ion electrode materials, a separate down-stream processing step is required for making carboncoated materials.

Carbon nanotubes (CNT) are also considered to be promising candidates for a high performance Lithium-ion anodes due to their high surface area and also their outstanding electrical and mechanical properties. However, the experimental results show only a 20-25% increase for reversible capacity compared to graphite electrode because of weak lithium adsorption to CNTs. To improve Li adsorption, researchers have used either chemical/plasma etching for making defects or fullerene (C60) encapsulation of single-walled carbon nanotubes for the creation of fullerene "peapods." In some cases, materials with a maximum capacity of ~1,000 mAh/g, corresponding to $Li_{2.7}C_6$, are produced. However, these methods require complex and high-temperature processing, often leading to deterioration in their performance.

The Future for Lithium-ion Energy Storage Materials

Emerging applications have steered Lithium-ion materials R&D in a new direction, which includes development of nanomaterial electrodes. Early versions of these nanomaterials are already beginning to appear in limited quantities in the marketplace, primarily in portable power tool applications. Within the next few years, Lithium-ion nanomaterials can also be expected to appear in automotive applications like PHEV and also in battery electrical energy storage systems. Existing applications like portable electronics, where Lithium-ion technology is fully entrenched, could also benefit from the implementation of nanomaterial electrodes that provide high-capacity at the desired powerlevels without compromising cycle life.

While adapting to the electrical performance requirements of the emerging applications, the materials and the processes used to make them have to be aware of the cost-sensitivity in the marketplace. Reduction in raw materials cost and consolidation of multiple steps in the process will be critical to any commercialization effort. Since Lithium-ion electrodes pack more energy and are in the proximity of non-aqueous electrolytes, they will also incur higher packaging costs at the cell-level and at the battery pack-level, compared to Nickel-Metal Hydride systems. Thermal cut-off devices in the cells and safety circuits and thermal management systems in the packs will ensure that the battery is well within its safe operating limit. In the long run, the performance premium offered by Lithium-ion batteries will more than off-set the added costs, particularly with economies of scale.

Mechanics of Stiff Thin Films of Controlled Wavy Geometry on Compliant Substrates for Stretchable Electronics

A thin film is a layer of material ranging from fractions of a nanometre (monolayer) to several micrometres in thickness. Electronic semiconductor devices and optical coatings are the main applications benefiting from thin-film construction. A familiar application of thin films is the household mirror, which typically has a thin metal coating on the back of a sheet of glass to form a reflective interface. The process of silvering was once commonly used to produce mirrors. A very-thin-film coating (less than about 50 nanometres thick) is used to produce two-way mirrors.

The performance of optical coatings (e.g., antireflective, or AR, coatings) are typically enhanced when the thin-film coating consists of multiple layers having varying thicknesses and refractive indices. Similarly, a periodic structure of alternating thin films of different materials may collectively form a so-called superlattice which exploits the phenomenon of quantum confinement by restricting electronic phenomena to two-dimensions.

Work is being done with ferromagnetic and ferroelectric thin films for use as computer memory. It is also being applied to pharmaceuticals, via thin-film drug delivery. Thin-films are used to produce thin-film batteries. Thin films are also used in dye-sensitized solar cells.

Ceramic thin films are in wide use. The relatively high hardness and inertness of ceramic materials make this type of thin coating of interest for protection of substrate materials against corrosion, oxidation and wear. In particular, the use of such coatings on cutting tools can extend the life of these items by several orders of magnitude.

Research is being done on a new class of thin-film inorganic oxide materials, called amorphous heavy-metal cation multicomponent oxides, which could be used to make transparent transistors that are inexpensive, stable, and environmentally benign.

Deposition

The act of applying a thin film to a surface is *thin-film deposition* – any technique for depositing a thin film of material onto a substrate or onto previously deposited layers. "Thin" is a relative term, but most deposition techniques control layer thickness within a few tens of nanometres. Molecular beam epitaxy allows a single layer of atoms to be deposited at a time.

It is useful in the manufacture of optics (for reflective, anti-reflective coatings or self-cleaning glass, for instance), electronics (layers of insulators, semiconductors, and conductors form integrated circuits), packaging (i.e., aluminium-coated PET film), and in contemporary art. Similar processes are sometimes used where thickness is not important: for instance, the purification of copper by electroplating, and the deposition of silicon and enriched uranium by a CVD-like process after gas-phase processing.

Deposition techniques fall into two broad categories, depending on whether the process is primarily chemical or physical.

Chemical Deposition

Here, a fluid precursor undergoes a chemical change at a solid surface, leaving a solid layer. An everyday example is the formation of soot on a cool object when it is placed inside a flame. Since the fluid surrounds the solid object, deposition happens on every surface, with little regard to direction; thin films from chemical deposition techniques tend to be *conformal*, rather than *directional*.

Chemical deposition is further categorized by the phase of the precursor:

- Plating relies on liquid precursors, often a solution of water with a salt of the metal to be deposited. Some plating processes are driven entirely by reagents in the solution (usually for noble metals), but by far the most commercially important process is electroplating. It was not commonly used in semiconductor processing for many years, but has seen a resurgence with more widespread use of chemical-mechanical polishing techniques.

- Chemical solution deposition (CSD) or Chemical bath deposition (CBD) uses a liquid precursor, usually a solution of organometallic powders dissolved in an organic solvent. This is a relatively inexpensive, simple thin film process that is able to produce stoichiometrically accurate crystalline phases. This technique is also known as the sol-gel method because the 'sol' (or solution) gradually evolves towards the formation of a gel-like diphasic system.
- Spin coating or spin casting, uses a liquid precursor, or sol-gel precursor deposited onto a smooth, flat substrate which is subsequently spun at a high velocity to centrifugally spread the solution over the substrate. The speed at which the solution is spun and the viscosity of the sol determine the ultimate thickness of the deposited film. Repeated depositions can be carried out to increase the thickness of films as desired. Thermal treatment is often carried out in order to crystallize the amorphous spin coated film. Such crystalline films can exhibit certain preferred orientations after crystallization on single crystal substrates.
- Chemical vapor deposition (CVD) generally uses a gas-phase precursor, often a halide or hydride of the element to be deposited. In the case of MOCVD, an organometallic gas is used. Commercial techniques often use very low pressures of precursor gas.
    - Plasma enhanced CVD (PECVD) uses an ionized vapor, or plasma, as a precursor. Unlike the soot example above, commercial PECVD relies on electromagnetic means (electric current, microwave excitation), rather than a chemical reaction, to produce a plasma.
- Atomic layer deposition (ALD) uses gaseous precursor to deposit conformal thin films one layer at a time. The process is split up into two half reactions, run in sequence and repeated for each layer, in order to ensure total layer saturation before beginning the next layer. Therefore, one reactant is deposited first, and then the second reactant is deposited, during which a chemical reaction occurs on the substrate, forming the desired composition. As a result of the stepwise, the process is slower than CVD, however it can be run at low temperatures, unlike CVD.

Physical Deposition

Physical deposition uses mechanical, electromechanical or thermodynamic means to produce a thin film of solid. An everyday example is the formation of frost. Since most engineering materials are held together by relatively high energies, and chemical reactions are not used to store these energies, commercial physical deposition systems tend to require a low-pressure vapor environment to function properly; most can be classified as physical vapor deposition (PVD).

The material to be deposited is placed in an energetic, entropic environment, so that particles of material escape its surface. Facing this source is a cooler surface which draws energy from these particles as they arrive, allowing them to form a solid layer. The whole system is kept in a vacuum deposition chamber, to allow the particles to travel as freely as possible. Since particles tend to follow a straight path, films deposited by physical means are commonly *directional*, rather than *conformal*.

Examples of physical deposition include:

- A thermal evaporator uses an electric resistance heater to melt the material and raise its vapor pressure to a useful range. This is done in a high vacuum, both to allow the vapor to reach the substrate without reacting with or scattering against other gas-phase atoms in the chamber, and reduce the incorporation of impurities from the residual gas in the vacuum chamber. Obviously, only materials with a much higher vapor pressure than the heating element can be deposited without contamination of the film. Molecular beam epitaxy is a particularly sophisticated form of thermal evaporation.
    - o An electron beam evaporator fires a high-energy beam from an electron gun to boil a small spot of material; since the heating is not uniform, lower vapor pressure materials can be deposited. The beam is usually bent through an angle of 270° in order to ensure that the gun filament is not directly exposed to the evaporant flux. Typical deposition rates for electron beam evaporation range from 1 to 10 nanometres per second.
    - o In molecular beam epitaxy (MBE), slow streams of an element can be directed at the substrate, so that material deposits one atomic layer at a time. Compounds such as gallium arsenide are usually deposited by repeatedly applying a layer of one element (i.e., gallium), then a layer of the other (i.e., arsenic), so that the process is chemical, as well as physical. The beam of material can be generated by either physical means (that is, by a furnace) or by a chemical reaction (chemical beam epitaxy).
- Sputtering relies on a plasma (usually a noble gas, such as argon) to knock material from a "target" a few atoms at a time. The target can be kept at a relatively low temperature, since the process is not one of evaporation, making this one of the most flexible deposition techniques. It is especially useful for compounds or mixtures, where different components would otherwise tend to evaporate at different rates. Note, sputtering's step coverage is more or less conformal. It is also widely used in the optical media. The manufacturing of all formats of CD, DVD, and BD are done with the help of this technique. It is a fast

technique and also it provides a good thickness control. Presently, nitrogen and oxygen gases are also being used in sputtering.

- Pulsed laser deposition systems work by an ablation process. Pulses of focused laser light vaporize the surface of the target material and convert it to plasma; this plasma usually reverts to a gas before it reaches the substrate.
- Cathodic arc deposition (arc-PVD) which is a kind of ion beam deposition where an electrical arc is created that literally blasts ions from the cathode. The arc has an extremely high power density resulting in a high level of ionization (30–100%), multiply charged ions, neutral particles, clusters and macro-particles (droplets). If a reactive gas is introduced during the evaporation process, dissociation, ionization and excitation can occur during interaction with the ion flux and a compound film will be deposited.
- Electrohydrodynamic deposition (electrospray deposition) is a relatively new process of thin film deposition. The liquid to be deposited, either in the form of nano-particle solution or simply a solution, is fed to a small capillary nozzle (usually metallic) which is connected to a high voltage. The substrate on which the film has to be deposited is connected to ground. Through the influence of electric field, the liquid coming out of the nozzle takes a conical shape (Taylor cone) and at the apex of the cone a thin jet emanates which disintegrates into very fine and small positively charged droplets under the influence of Rayleigh charge limit. The droplets keep getting smaller and smaller and ultimately get deposited on the substrate as a uniform thin layer.

Thin-Film Photovoltaic Cells

Thin-film technologies are also being developed as a means of substantially reducing the cost of photovoltaic (PV) systems. The rationale for this is that thin-film modules are cheaper to manufacture owing to their reduced material costs, energy costs, handling costs and capital costs. This is especially represented in the use of printed electronics (roll-to-roll) processes.

Thin films belong to the second and third photovoltaic cell generations.

Plasmonic Solar Cell

Plasmonic solar cells (PSC) are a class of photovoltaic devices that convert light into electricity by using plasmons. PSCs are a type of thin-film SC which are typically 1-2μm thick. They can use substrates which are cheaper than silicon, such as glass, plastic or steel. The biggest problem for thin film solar cells is that they don't absorb as much light as the current solar cells. Methods for trapping light on the surface, or in the SC are crucial in order to make thin film SCs viable. One method which has been explored over the past few years is to scatter light using metal nanoparticles excited at their surface plasmon resonance. This allows light to be absorbed more directly without the relatively thick additional layer required in other types of thin-film solar cells.

History

People: There have been quite a few pioneers working with plasmonic solar cells. One of the main focuses has been on improving the thin film SC through the use of metal nanoparticles distributed on the surface. It has been found that the Raman scattering can be increased by order of magnitude when using metal nanoparticles. The increased Raman scattering provides more photons to become available to excite surface plasmons which cause electrons to be excited and travel through the thin film SC to create a current. The list below shows a few of research which has been done to improve PSCs.

- Stuart and Hall: Photocurrent enhancement by 18x with 165 nm SOI photodetector with wavelength of 800 nm using silver nanoparticles used for scattering and absorption of light.
- Schaadt: Gold nanoparticles used for scattering and absorption of light on doped silicon obtaining 80% enhancements with 500 nm wavelength.
- Derkacs: Gold nanoparticles on thin-film silicon gaining 8% on conversion efficiency.
- Pillai: Silver particles on SOI obtaining 33% photocurrent increase.
- Stenzel: Enhancements in photocurrent by a factor of 2.7 for ITO-copper phthalocyanine-indium structures.
- Westphalen: Enhancement for silver clusters incorporated into ITO and zinc phthalocyanine solar cells.
- Rand: Enhanced efficiencies for ultra thin film organic solar cells due to 5 nm diameter silver nanoparticles.
- Brown: Enhanced photocurrent and efficiencies in dye-sensitized solar cells incorporating metal-insulator core-shell nanoparticle geometries.

Devices: There are currently three different generations of SCs. The first generation (those in the market today) are made with crystalline semiconductor wafers, typically silicon. These are the SCs everybody thinks of when they hear "Solar Cell".

Current SCs trap light by creating pyramids on the surface which have dimensions bigger than most thin film SCs. Making the surface of the substrate rough (typically by growing SnO_2 or ZnO on surface) with dimensions on the order of the incoming wavelengths and depositing the SC on top has been explored. This method increases the photocurrent, but the thin film SC would then have poor material quality.

The second generation SCs are based on thin film technologies such as those presented here. These SCs focus on lowering the amount of material used as well as increasing the energy production. Third generation SCs are currently being researched. They focus on reducing the cost of the second generation SCs. The third generation SCs are discussed in more detail under recent advancement.

Design: The design for a PSC varies depending on the method being used to trap and scatter light across the surface and through the material.

Metal Nanoparticle Plasmonic Solar Cell

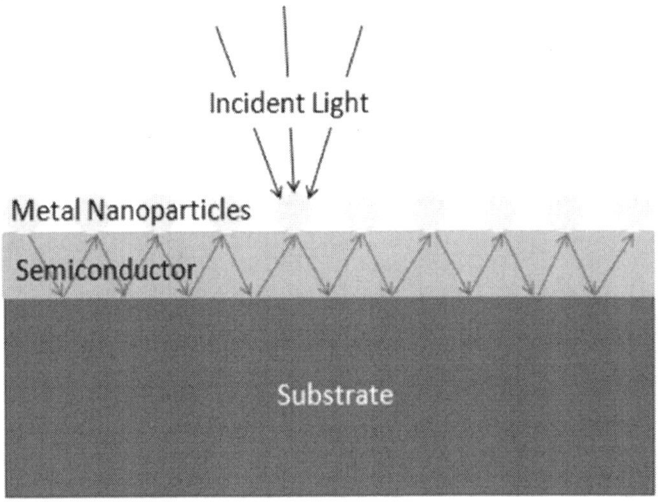

Figure: *PSC using metal nanoparticles.*

A common design is to deposit metal nanoparticles on the top surface of the thin film SC. When light hits these metal nanoparticles at their surface plasmon resonance, the light is scattered in many different directions. This allows light to travel along the SC and bounce between the substrate and the nanoparticles enabling the SC to absorb more light.

Metal Film Plasmonic Solar Cell

Other methods utilising surface plasmons for harvesting solar energy are available. One other type of structure is to have a thin film of silicon and a thin layer of metal deposited on the lower surface. The light will travel through the silicon and generate surface plasmons on the interface of the silicon and metal. This generates electric fields inside of the silicon since electric fields do not travel very far into metals. If the electric field is strong enough, electrons can be moved and collected to produce a photocurrent. The thin film of metal in this design must have nanometre sized grooves which act as waveguides for the incoming light in order to excite as many photons in the silicon thin film as possible.

Basic Principles

General:

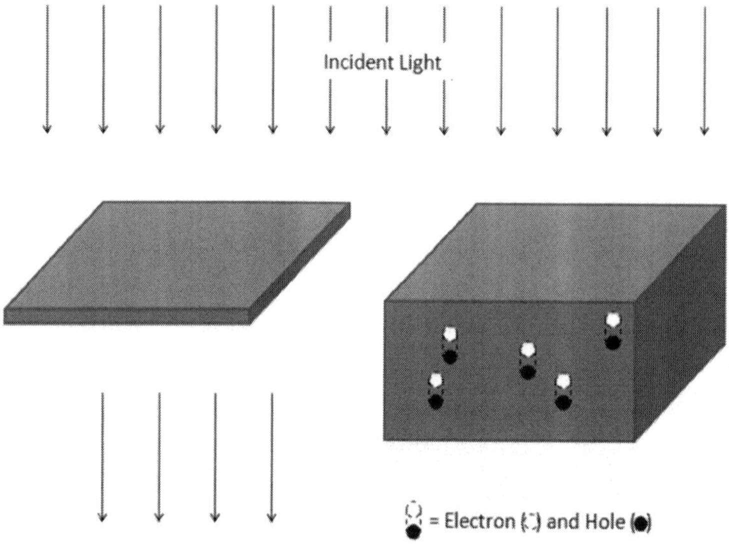

Figure: *Thin film SC (left) and Typical SC (right).*

When a photon is excited in the substrate of a SC, an electron and hole are separated. Once the electrons and holes are separated, they will want to recombine since they are of opposite charge. If the electrons can be collected prior to this happening then the SC is pretty efficient. The way to collect the electrons quickly would be to make the conducting material very thin. If the surface is made very thin, there will be less light absorbed by the device. A thick device absorbs more light.

Scattering and Absorption

The basic principles for the functioning of plasmonic solar cells include scattering and absorption of light due to the deposition of metal nanoparticles. Silicon does not absorb light very well. For this reason, more light needs to be scattered across the surface in order to increase the absorption. It has been found that metal nanoparticles help to scatter the incoming light across the surface of the silicon substrate.

The equations that govern the scattering and absorption of light can be shown as:

- $C_{scat} = \frac{1}{6\pi}\left(\frac{2\pi}{\lambda}\right)^4 |\alpha|^2$

This shows the scattering of light for particles which have diameters below the wavelength of light.

- $C_{abs} = \frac{2\pi}{\lambda} Im[\alpha]$

This shows the absorption for a point dipole model.

- $\alpha = 3V\left[\frac{\epsilon_p / \epsilon_m - 1}{\epsilon_p / \epsilon_m + 2}\right]$

This is the polarizability of the particle. V is the particle volume. ϵ_p is the dielectric function of the particle. ϵ_m is the dielectric function of the embedding medium. When $\epsilon_p = -2\epsilon_m$ the polarizability of the particle becomes large. This polarizability value is known as the surface plasmon resonance. The dielectric function for metals with low absorption can be defined as:

- $\epsilon = 1 - \frac{\omega_p^2}{\omega^2 + i\gamma\omega}$

In the previous equation, ω_p is the bulk plasma frequency. This is defined as:

- $\omega_p^2 = Ne^2 / m\epsilon_0$

N is the density of free electrons, e is the electronic charge and m is the effective mass of an electron. ϵ_0 is the dielectric constant of free space. The equation for the surface plasmon resonance in free space can therefore be represented by:

- $\alpha = 3V \frac{\omega_p^2}{\omega_p^2 - 3\omega^2 - i\gamma\omega}$

Many of the plasmonic solar cells use nanoparticles to enhance the scattering of light. These nanoparticles take the shape of spheres, and therefore the surface plasmon resonance frequency for spheres is desirable. By solving the previous equations, the surface plasmon resonance frequency for a sphere in free space can be shown as:

- $\omega_{sp} = \sqrt{3}\omega_p$

As an example, at the surface plasmon resonance for a silver nanoparticle, the scattering cross-section is about 10x the cross-section of the nanoparticle. The goal of the nanoparticles is to trap light on the surface of the SC. The absorption of light is not important for the nanoparticle, rather, it is important for the SC. One would think that if the nanoparticle is increased in size, then the scattering cross-section becomes larger. This is true, however, when compared with the size of the nanoparticle, the ratio ($\frac{CS_{scat}}{CS_{particle}}$) is reduced. Particles with a large scattering cross section tend to have a broader plasmon resonance range.

Wavelength Dependence

Surface plasmon resonance mainly depends on the density of free electrons in the particle. The order of densities of electrons for different metals is shown below along with the type of light which corresponds to the resonance.

- Aluminum - Ultra-violet
- Silver - Ultra-violet
- Gold - Visible
- Copper - Visible

If the dielectric constant for the embedding medium is varied, the resonant frequency can be shifted. Higher indexes of refraction will lead to a longer wavelength frequency.

Light Trapping

The metal nanoparticles are deposited at a distance from the substrate in order to trap the light between the substrate and the particles. The particles are embedded in a material on top of the substrate. The material is typically a dielectric, such as silicon or silicon nitride. When performing experiment and simulations on the amount of light scattered into the substrate due to the distance between the particle and substrate, air is used as the embedding material as a reference. It has been found that the amount of light radiated into the substrate decreases with distance from the substrate. This means that nanoparticles on the surface are desirable for radiating light into the substrate, but if there is no distance between the particle and substrate, then the light is not trapped and more light escapes.

The surface plasmons are the excitations of the conduction electrons at the interface of metal and the dielectric. Metallic nanoparticles can be used to couple and trap freely propagating plane waves into the semiconductor thin film layer. Light can be folded into the absorbing layer to increase the absorption. The localized surface plasmons in metal nanoparticles and the surface plasmon polaritons at the interface of metal and semiconductor are of interest in the current research. In recent reported papers, the shape and size of the metal nanoparticles are key factors to determine the incoupling efficiency. The smaller particles have larger incoupling efficiency due to the enhanced near-field coupling. However, very small particles suffer from large ohmic losses.

Metal Film

As light is incident upon the surface of the metal film, it excites surface plasmons. The surface plasmon frequency is specific for the material, but through the use of gratings on the surface of the film, different frequencies can be obtained. The surface plasmons are also preserved through the use of waveguides as they make the surface plasmons easier to travel on the surface and the losses due to resistance and radiation are minimized. The electric field generated by the surface plasmons influences the electrons to travel toward the collecting substrate.

Materials

First Generation	*Second Generation*	*Third Generation*
Single-crystal silicon	CuInSe2	Gallium Indium Phosphide
Multicrystalline silicon	amorphous silicon	Gallium Indium Arsenide
Polycrystalline silicon	thin film crystalline Si	Germanium

Applications

The applications for plasmonic solar cells are endless. The need for cheaper and more efficient solar cells is huge. In order for solar cells to be considered cost effective, they need to provide energy for a smaller price than that of traditional power sources such as coal and gasoline. The movement toward a more green world has helped to spark research in the area of plasmonic solar cells. Currently, solar cells cannot exceed efficiencies of about 30% (First Generation). With new technologies (Third Generation), efficiencies of up to 40-60% can be expected. With a reduction of materials through the use of thin film technology (Second Generation), prices can be driven lower.

Space

Certain applications for plasmonic solar cells would be for space exploration vehicles. A main contribution for this would be the reduced weight of the solar cells. An external fuel source would also not be needed if enough power could be generated from the solar cells. This would drastically help to reduce the weight as well.

Rural

Solar cells have a great potential to help rural electrification. An estimated two million villages near the equator have limited access to electricity and fossil fuels and that approximately 25% of people in the world do not have access to electricity. When the cost of extending power grids, running rural electricity and using diesel generators is compared with the cost of solar cells, many times the solar cells win. If the efficiency and cost of the current solar cell technology is decreased even further, then many rural communities and villages around the world could obtain electricity when current methods are out of the question. Specific applications for rural communities would be water pumping systems, residential electric supply and street lights. A particularly interesting application would be for health systems in countries where motorized vehicles are not overly abundant. Solar cells could be used to provide the power to refrigerate medications in coolers during transport.

Solar cells could also provide power to lighthouses, buoys, or even battleships out in the ocean. Industrial companies could use them to power telecommunications systems or monitoring and control systems along pipelines or other system.

High Power

If the solar cells could be produced on a large scale and be cost effective then entire power stations could be built in order to provide power to the electrical grids. With a reduction in size, they could be implemented on both commercial and residential buildings with a much smaller footprint. They might not even seem like an eyesore. Other areas are in hybrid systems. The solar cells could help to power high consumption devices such as automobiles in order to reduce the amount of fossil fuels used and to help improve the environmental conditions of the earth.

Low Power

One application which has not been mentioned is consumer electronics. Essentially, solar cells could be used to replace batteries for low power electronics. This would save everyone a lot of money and it would also help to reduce the amount of waste going into landfills.

Recent Advancements

Choice of Plasmonic Metal Nanoparticles

Proper choice of plasmatic metal nanoparticles is crucial for the maximum light absorption in the active layer. Front surface located nanoparticles Ag and Au are the most widely used materials due to their surface plasmon resonances located in the visible range and therefore interact more strongly with the peak solar intensity. However, such noble metal nanoparticles always introduce reduced light coupling into Si at the short wavelengths below the surface plasmon resonance due to the detrimental Fano effect, i.e. the destructive interference between the scattered and unscattered light. Moreover, the noble metal nanoparticles are impractical to implement for large-scale solar cell manufacture due to their high cost and scarcity in earth crest. Recently, Zhang et al. have demonstrated the low cost and earth abundant materials Al nanoparticles to be able to outperform the widely used Ag and Au nanoparticles. Al nanoparticles, with their surface plasmon resonances located in the UV region below the desired solar spectrum edge at 300 nm, can avoid the reduction and introduce extra enhancement in the shorter wavelength range.

Light Trapping

As discussed earlier, being able to concentrate and scatter light across the surface of the plasmonic solar cell will help to increase efficiencies. Recently, research at Sandia National Laboratories has discovered a photonic waveguide which collects light at a certain wavelength and traps it within the structure. This new structure can contain 95% of the light that enters it compared to 30% for other traditional waveguides. It can also direct the light within one wavelength which is ten times greater than traditional waveguides. The wavelength this device captures can be selected by changing the structure of the lattice which comprises the structure. If this structure is used to trap light and keep it in the structure until the solar cell can absorb it, the efficiency of the solar cell could be increased dramatically.

Absorption

Another recent advancement in plasmonic solar cells is using other methods to aid in the absorption of light. One way being researched is the use of metal wires on top of the substrate to scatter the light. This would help by utilising a larger area of the surface of the solar cell for light scattering and absorption. The danger in using lines instead of dots would be creating a reflective layer which would reject light from the system. This is very undesirable for solar cells. This would be very similar to the thin metal film approach, but it also utilises the scattering effect of the nanoparticles.

Third Generation Solar Cells

The goal of third generation solar cells is to increase the efficiency using second generation solar cells (thin film) and using materials that are found abundantly on earth. This has also been a goal of the thin film solar cells. With the use of common and safe materials, third generation solar cells should be able to be manufactured in mass quantities further reducing the costs. The initial costs would be high in order to produce the manufacturing processes, but after that they should be cheap. The way third generation solar cells will be able to improve efficiency is to absorb a wider range of frequencies. The current thin film technology has been limited to one frequency due to the use of single band gap devices.

Multiple Energy Levels

The idea for multiple energy level solar cells is to basically stack thin film solar cells on top of each other. Each thin film solar cell would have a different band gap which means that if part of the solar spectrum was not absorbed by the first cell then the one just below would be able to absorb part of the spectrum. These can be stacked and an optimal band gap can be used for each cell in order to produce the maximum amount of power. Options for how each cell is connected are available, such as serial or parallel. The serial connection is desired because the output of the solar cell would just be two leads.

The lattice structure in each of the thin film cells needs to be the same. If it is not then there will be losses. The processes used for depositing the layers are complex. They include Molecular Beam Epitaxy and Metal Organic Vapour Phase Epitaxy. The current efficiency record is made with this process but doesn't have exact matching lattice constants. The losses due to this are not as effective because the differences in lattices allows for more optimal band gap material for the first two cells. This type of cell is expected to be able to be 50% efficient.

Lower quality materials that use cheaper deposition processes are being researched as well. These devices are not as efficient, but the price, size and power combined allow them to be just as cost effective. Since the processes are simpler and the materials are more readily available, the mass production of these devices is more economical.

Hot Carrier Cells

A problem with solar cells is that the high energy photons that hit the surface are converted to heat. This is a loss for the cell because the incoming photons are not converted into usable energy. The idea behind the hot carrier cell is to utilise some of that incoming energy which is converted to heat. If the electrons and holes can be collected while hot, a higher voltage can be obtained from the cell. The problem with doing this is that the contacts which collect the electrons and holes will cool the material. Thus far, keeping the contacts from cooling the cell has been theoretical. Another way of improving the efficiency of the solar cell using the heat generated is to have a cell which allows lower energy photons to excite electron and hole pairs. This requires a small bandgap. Using a selective contact, the lower energy electrons and holes can be collected while allowing the higher energy ones to continue moving through the cell. The selective contacts are made using a double barrier resonant tunneling structure. The carriers are cooled which they scatter with phonons. If a material with a large bandgap of phonons then the carriers will carry more of the heat to the contact and it won't be lost in the lattice structure. One material which has a large bandgap of phonons is indium nitride. The hot carrier cells are in their infancy but are beginning to move toward the experimental stage.

4

Thin-Film Batteries

Thin-film printing technology is being used to apply solid-state lithium polymers to a variety of substrates to create unique batteries for specialised applications. Thin-film batteries can be deposited directly onto chips or chip packages in any shape or size. Flexible batteries can be made by printing onto plastic, thin metal foil, or paper.

Thin film lithium ion batteries are similar to conventional lithium-ion batteries, but they are composed of thinner materials, some only nanometres or micrometres thick, which allow the finished battery to be just millimetres thick. They have been developed primarily within the last decade. These batteries consist of a substrate, electrolyte, current collector, anode, cathode, and a separator.

There has been much research into the determination of the most effective components for this type of battery. It has been shown recently that even ordinary printer paper can be used as a charge separator and a substrate. These thin film batteries are an improvement on the common secondary, or rechargeable, lithium ion batteries in some ways.

The batteries exhibit the same voltage and current as their bulky counterparts, but their dimensions allow for making electronic devices like cell phones, laptops and even implantable medical devices smaller and also reduce the weight of portable battery powered devices because of their high energy density.

These batteries can be formed into any shape and can be stacked thus further reducing the space needed.

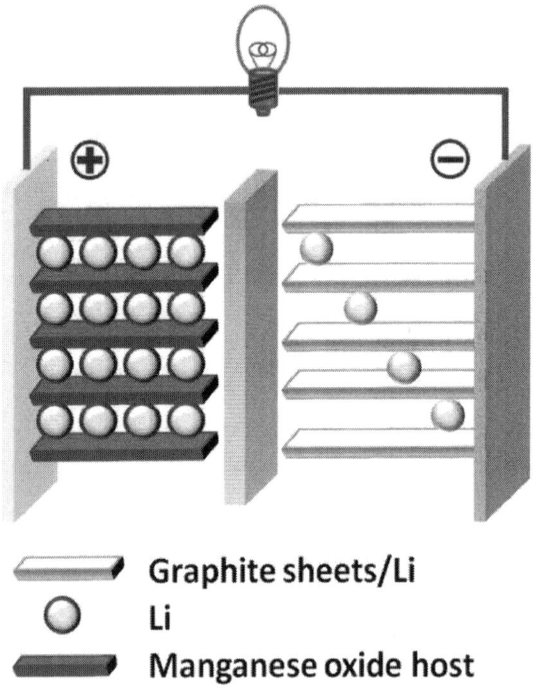

Figure: *Lithium Ion Battery*

Lithium-ion batteries are different from disposable lithium batteries because they have the ability to be recharged. They have high energy density and last as long as other battery technologies. Inside the battery, positive lithium ions flow through the electrolyte from the anode to the cathode while the battery is being discharged. When recharging the battery, lithium ions move back to the anode.

The Li-ion battery design is effective for supplying portable devices with larger power requirements.

As a mobile, technology driven society we rely on portable electronics, which require thinner batteries for power. Research into thin film batteries has increased in recent years to accommodate this demand.

Components of Thin Film Battery

Cathode Materials

Cathode materials in thin film lithium ion batteries are the same of what is seen in classical lithium ion batteries. They are normally metal oxides that are deposited as a film by various methods.

Metal oxide materials are shown below as well as their relative specific capacities (Λ), open circuit voltages (V_{oc}), and energy densities (D_E).

Material Ratings

	Ë(Ah/kg)	V_{OC}(V)	D_E(Wh/kg)
$LiCoO_2$	145	4	580
$LiMn_2O_4$	148	4	592
$LiFePO_4$	170	3.4	578

Energy Density

$$D_E = \Lambda\ V_{OC}$$

Λ: capacity (mAh/g)

V_{OC}: Open circuit potential

Deposition Methods for Cathode Materials

There are various methods being used to deposit thin film cathode materials onto the current collector.

Pulsed Laser Deposition (PLD)

In Pulsed Laser Deposition, materials are fabricated by controlling parameters such as laser energy and fluence, substrate temperature, background pressure, and target-substrate distance.

Magnetron Sputtering

In Magnetron Sputtering the substrate is cooled for deposition.

Chemical Vapor Deposition (CVD)

In Chemical Vapor Deposition, volatile precursor materials are deposited onto a substrate material.

Sol-Gel Processing

Sol-gel processing allows for homogeneous mixing of precursor materials at the atomic level.

Electrolyte

The greatest difference between classical lithium ion batteries and thin, flexible, lithium ion batteries is in the electrolyte material used. Progress in lithium ion batteries relies as much on improvements in the electrolyte as it does in the electrode materials, as the electrolyte plays a major role in safe battery operation. The concept of thin film lithium ion batteries was increasingly motivated by manufacturing advantages presented by the polymer technology for their use as electrolytes. LiPON, lithium phosphorus oxynitride, is an amorphous glassy material used as an electrolyte material in thin film flexible batteries. Layers of LiPON are deposited over the cathode material at ambient temperatures by RF magnetron sputtering. This forms the solid electrolyte used for ion conduction between anode and cathode. Solid polymer electrolytes offer several advantages in comparison to a classical liquid lithium ion battery. Rather than having separate components of electrolyte, binder, and separator, these solid electrolytes can act as all three. This increases the overall energy density of the assembled battery because the constituents of the entire cell are more tightly packed.

Separator Material

Separator materials in lithium ion batteries must not block the transport of lithium ions while preventing the physical contact of the anode and cathode materials, e.g. short-circuiting. In a liquid cell, this separator would be a porous glass or polymer mesh that allows ion transport via the liquid electrolyte through the pores, but keeps the electrodes from contacting and shorting. However, in a thin film battery the electrolyte is a solid, which conveniently satisfies both the ion transportation and the physical separation requirements without the need for a dedicated separator.

Current Collector

Current collectors in thin film batteries must be flexible, have high surface area, and be cost-effective. Silver nanowires with improved surface area and loading weight have been shown to work as a current collector in these battery systems, but still are not as cost-effective as desired. Extending graphite technology to lithium ion batteries, solution processed carbon nanotubes (CNT) films are being looked into for use as both the current collector and anode material. CNTs have the ability to intercalate lithium and maintain high operating voltages, all with low mass loading and flexibility.

Advantages and Challenges

Thin film lithium ion batteries offer improved performance by having a higher average output voltage, lighter weights thus higher energy density, and longer cycling life than typical rechargeable batteries. In the thin film lithium ion battery, both electrodes are capable of reversible lithium insertion, thus forming a Li-ion transfer cell. Li-ion transfer cells are the most promising systems for satisfying the demand of high specific energy and high power. In order to construct a thin film battery it is necessary to fabricate all the battery components, as an anode, a solid electrolyte, a cathode and current leads into multi-layered thin films by suitable technologies.

In a thin film based system, the electrolyte is normally a solid electrolyte, capable of conforming to the shape of the battery. This is in contrast to classical lithium ion batteries, which normally have liquid electrolyte material. Liquid electrolytes can be challenging to utilise if they are not compatible with the separator. Also liquid electrolytes in general call for an increase in the overall volume of the battery, which is not ideal for designing a system that has high energy density. Additionally, in a thin film flexible Li-ion battery, the electrolyte, which is normally polymer-based, can act as the electrolyte, separator, and binder material. This provides the ability to have flexible systems since the issue of electrolyte leakage is circumvented. Finally, solid systems can be packed together tightly which affords an increase in energy density when compared to classical lithium ion batteries.

Separator materials in lithium ion batteries must have the ability to transport ions through their porous membranes while maintaining a physical separation between the anode and cathode materials in order to prevent short-circuiting. Furthermore, the separator must be resistant to degradation during the battery's operation. In a thin film Li-ion battery, the separator must be a thin and flexible solid. Typically today, this material is a polymer-based material. Since thin film batteries are made of all solid materials, allows one to use simpler separator materials in these systems such as Xerox paper rather than in liquid based Li-ion batteries.

Scientific Development

Development of thin solid state batteries allows for roll to roll type production of batteries to decrease production costs. Solid-state batteries can also afford increased energy density due to decrease in overall device weight, while the flexible nature allows for novel battery design and easier incorporation into electronics. Development is still required in cathode materials which will resist capacity reduction due to cycling.

Prior Technology	***Replacement Technology***	***Result***
Solution based electrolyte	Solid state electrolyte	Increased safety and cycle life
Polymer separators	Paper separator	Decreased cost increased rate of ion conduction
Metallic current collectors	Carbon nanotube current collectors	Decreased device weight, increased energy density
Graphite anode	Carbon nanotube anode	Decreased device complexity

Applications

The advancements made to the thin film lithium ion battery have allowed for many potential applications. The majority of these applications are aimed at improving the currently available consumer and medical products. Thin film lithium ion batteries can be used to make thinner portable electronics, because the thickness of the battery required to operate the device can be reduced greatly. These batteries have the ability to be an integral part of implantable medical devices, such as defibrillators and neural stimulators, "smart" cards, radio frequency identification, or RFID, tags and wireless sensors. They can also serve as a way to store energy collected from solar cells or other harvesting devices. Each of these applications is possible because of the flexibility in the size and shape of the batteries. The size of these devices does not have to revolve around the size of the space needed for the battery anymore. The thin film batteries can be attached to the inside of the casing or in some other convenient way. The opportunities in which to use this type of batteries are endless.

Solar Cell Storage Devices

The thin film lithium ion battery can serve as a storage device for the energy collected from a solar cell. These batteries can be made to have a low self discharge rate, which means that these batteries can be stored for long periods of time without a major loss of the energy that was used to charge it. These fully charged batteries could then be used to power some or all of the other potential applications listed below.

Smart Cards

Smart cards have the same size as a credit card, but they contain a microchip that can be used to access information, give authorization, or process an application. These cards can go through harsh production conditions, with temperatures in the range of 130 to 150 °C, in order to complete the high temperature, high pressure lamination processes. These conditions can cause other batteries to fail because of degassing or degradation of organic components within the battery. Thin film lithium ion batteries have been shown to withstand temperatures of -40 to 150 °C. This use of thin film lithium ion batteries is hopeful for other extreme temperature applications.

RFID Tags

Radio Frequency Identification (RFID) tags can be used in many different applications. These tags can be used in packaging, inventory control, used to verify authenticity and even allow or deny access to something. These ID tags can even have other integrated sensors to allow for the physical environment to be monitored, such as temperature or shock during travel or shipping. Also, the distance required to read the information in the tag depends on the strength of the battery.

The farther away you want to be able to read the information, the stronger the output will have to be and thus the greater the power supply to accomplish this output. As these tags get more and more complex, the battery requirements will need to keep up. Thin film lithium ion batteries have shown that they can fit into the designs of the tags because of the flexibility of the battery in size and shape and are sufficiently powerful enough to accomplish the goals of the tag. Low cost production methods, like roll to roll lamination, of these batteries may even allow for this kind of RFID technology to be implemented in disposable applications.

Implantable Medical Devices

Thin films of $LiCoO_2$ have been synthesized in which the strongest x ray reflection is either weak or missing, indicating a high degree of preferred orientation. Thin film solid state batteries with these textured cathode films can deliver practical capacities at high current densities. For example, for one of the cells 70% of the maximum capacity between 4.2 V and 3 V (approximately 0.2 mAh/cm^2) was delivered at a current of 2 mA/cm^2. When cycled at rates of 0.1 mA/cm^2, the capacity loss was 0.001%/cycle or less. The reliability and performance of Li $LiCoO_2$ thin-film batteries make them attractive for application in implantable devices such as neural stimulators, pacemakers, and defibrillators.

Implantable medical devices require batteries that can deliver a steady, reliable power source for as long as possible. These applications call for a battery that has a low self-discharge rate, for when it's not in use, and a high power rate, for when it needs to be used, especially in the case of an implantable defibrillator. Also, users of the product will want a battery that can go through many cycles, so these devices will not have to be replaced or serviced often. Thin film lithium ion batteries have the ability to meet these requirements. The advancement from a liquid to a solid electrolyte has allowed these batteries to take almost any shape without the worry of leaking, and it has been shown that certain types of thin film rechargeable lithium batteries can last for around 50,000 cycles. Another advantage to these thin film batteries is that they can be arranged in series to give a larger voltage equal to the sum of the individual battery voltages. This fact can be used in reducing the "footprint" of the battery, or the size of the space needed for the battery, in the design of a device.

Wireless Sensors

Wireless sensors need to be in use for the duration of their application, whether that may be in package shipping or in the detection of some unwanted compound, or controlling inventory in a warehouse. If the wireless sensor cannot transmit its data due to low or no battery power, the consequences could potentially be severe based on the application. Also, the wireless sensor must be adaptable to each application. Therefore the battery must be able to fit within the designed sensor. This means that the desired battery for these devices must be long-lasting, size specific, low cost, if they are going to be used in disposable technologies, and must meet the requirements of the data collection and transmission processes. Once again, thin film lithium ion batteries have shown the ability to meet all of these requirements.

Stretchable Interconnects

In telecommunications, interconnection is the physical linking of a carrier's network with equipment or facilities not belonging to that network. The term may refer to a connection between a carrier's facilities and the equipment belonging to its customer, or to a connection between two (or more) carriers.

In United States regulatory law, interconnection is specifically defined (47 C.F.R. 51.5) as "the linking of two networks for the mutual exchange of traffic."

One of the primary tools used by regulators to introduce competition in telecommunications markets has been to impose interconnection requirements on dominant carriers.

History

United States: Under the Bell System monopoly (post Communications Act of 1934), the Bell System owned the phones and did not allow interconnection, either of separate phones (or other terminal equipment) or of other networks; a popular saying was "Ma Bell has you by the calls".

This began to change in the landmark case Hush-A-Phone v. United States [1956], which allowed some non-Bell owned equipment to be connected to the network, and was followed by a number of other cases, regulatory decisions, and legislation that led to the transformation of the American long distance telephone industry from a monopoly to a competitive business.

This further changed in FCC's *Carterfone* decision in 1968, which required the Bell System companies to permit interconnection by radio-telephone operators.

Today the standard electrical connector for interconnection in the US, and much of the world, is the registered jack family of standards, especially RJ11. This was introduced by the Bell System in the 1970s, following a 1976 FCC order. Since then, it has gained popularity worldwide, and is a de facto international standard.

Europe

Outside of the U.S., Interconnection or "Interconnect regimes" also take into account the associated commercial arrangements. As an example of the use of commercial arrangements, the focus by the EU has been on "encouraging" incumbents to offer bundles of network features that will enable competitors to provide services that compete directly with the incumbent. Further the interconnect regime decided upon by the regulator has a major impact on the development/rate of growth of market segments. According to Source8 (an EU based consultancy) two examples from the UK of this are:

- The decision about revenue sharing on local rate numbers was a contributory factor in the explosive growth in dial internet.
- The asynchronous reciprocity that exists between fixed and mobile termination rates.

Design for Reliability of Multi-Layer Thin Film Stretchable Interconnects

Most electronic systems that power our digital life are inflexible and flat. Rigid electronic designs work for our computers and phones but not for our bodies. Humans are soft and curved. Electronic systems capable of bending, twisting, and stretching have great potential for applications in which conventional, stiff semiconductor microelectronics would not suffice. One promising area for conformable electronics is in biomedical applications such as wearable and implantable electronic systems.

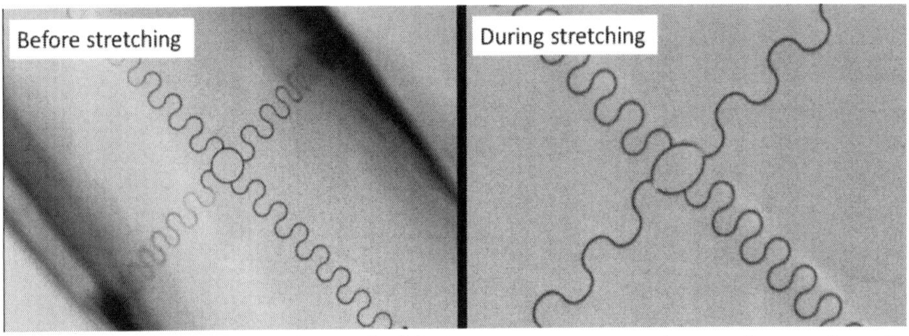

Figure: *Multiple layers of stretchable interconnects crossing at the strain relief structure described below.*

Previous stretchable interconnect designs are composed of only a single metal layer. Due to the nature of this single metal layer design, these stretchable electronic systems have large, sparse layouts. However, as the system gets more complex, single metal layer interconnects become impractical. Multi-layered stretchable interconnects allow for smaller, more intricate, and more practical systems and can be produced through conventional bottom-up micro-fabrication processes. When multiple meandering interconnects intersect, however, the junctions of the interconnects are subject to the specific orientation of each structure, complicating the design, risking the integrity of the small junction areas, and resulting in complex mechanics during stretching.

A circular structure to relieve strain at the junctions between multiple metal interconnect layers is proposed in the paper. In the proposed strain relief structure design, meandering horseshoe patterned metal interconnects contain circular rings that are stacked to allow the interconnects to overlap one another. While the proposed strain relief structure is a universal design that can be applied to any number of systems with multiple interconnecting layers, structures composed of two metal layers are examined in this study.

A comprehensive investigation into the deformation behaviour and failure mechanisms of the structures was carried out through both numerical and experimental analysis. Numerical analysis indicates that elongations of up to 20% cause no plastic strain in the structure, allowing it to operate indefinitely.

Simulations show that the crests of the horseshoe interconnects are the regions with the greatest strain and that the structure will ultimately fail at one of these crests.

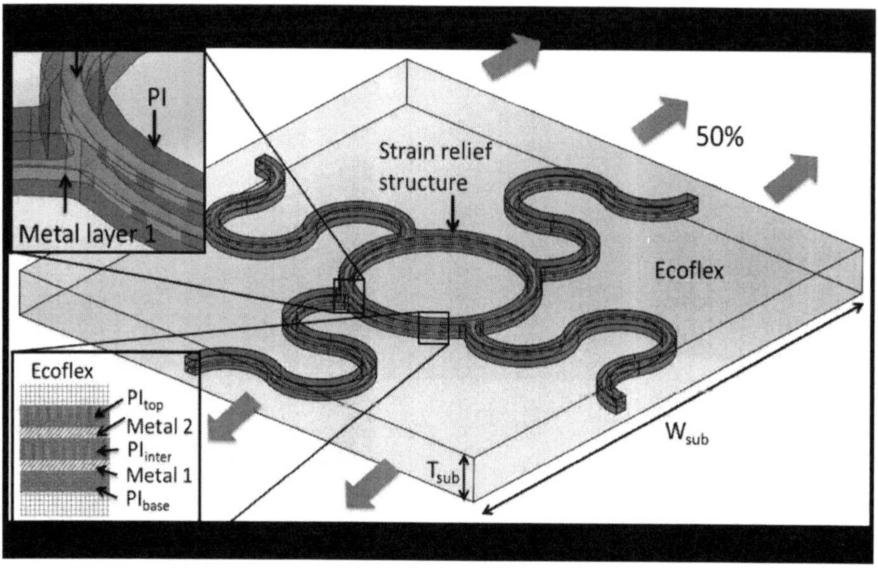

Figure: *Strain relief structure design with two layered intersecting interconnects.*

Results of electromechanical testing show that the strain relief structure can stretch up to 285% of its initial length prior to failure. Initial results from fatigue life testing of the structures has demonstrated that they are able to withstand more than one million cycles of 100% elongation at a 200% per second strain rate. Fatigue life testing also verified the numerical simulations, indicating that the strain relief structure effectively dissipates strain from the interconnect junctions to the crests of the horseshoes. Both the simulation and experimental results show that this multi-layer strain relief structure for stretchable electronic systems is a durable and highly promising design with significant implications for the future of conformable electronics.

Materials for Stretchable Electronics

Electronics can be made on elastically stretchable "skin." Such skins conform to irregularly curved surfaces and carry arrays of thinfilm devices and integrated circuits. Stretchable circuits are made of diverse materials that span more than orders of magnitude in elastic modulus.

These components and techniques provide the tools for creating new concepts in biocompatible circuits that conform to and stretch with living tissue. They enable wireless energy transfer via stretchable antennas, stretchable solar cells that convert sunlight to electricity, supercapacitors, and batteries that store energy in stretchable electronic devices.

Stretchable Electronics, the Newest Class of Large-Area Electronics

Stretchable electronics is the newest class of large-area electronics. Quite literally large-area electronics has become a success story: flat panel displays are manufactured at the rate of 100 and solar cells at 200 square kilometres a year. Many of these products are made with thin films. Liquid-crystal displays1 are driven by active matrices of amorphous silicon transistors, and 10% of all solar cells are made of amorphous silicon3 or chalcogenide films. Like any other stiff material, circuits become flexible and rollable when their thickness is reduced to 1/1000 of the desired radius of curvature. Thinfilm circuits are made on flat surfaces by standard microfabrication techniques. When made on plastic substrates and with plastically deformable interconnects, they can be shaped to surfaces that need expansion out of the plane, for instance spherical caps. On plastic substrates, this deformation is permanent. Elastomeric substrates and elastic interconnects let circuits go a shape beyond: to reversible deformation and near-arbitrary dimensions. Sizes and shapes of elastomeric circuits can be changed reversibly by applying mechanical force, by gas pressure, or by application of an electric field.

Now we can make electronic skin, conformable sensors and displays, electronic biointerfaces, electronic muscles, and energy harvesters. The latest progression is circuits made on a biodegradable substrate, which disappears to leave the bare circuit conforming to a living organism; the goal is a circuit that envelopes the heart fully to sense and stimulate it with high spatial resolution while expanding and shrinking with the heartbeat. Stretchable electronics is the research frontier of large-area electronics.

A materials scientist first will notice the elastic, soft nature of experimental stretchable electronics: a polymeric electroactuator, a wraparound pressure sensing skin, or a neural interface in the folds of the cerebral cortex. Stretchable electronics are fabricated with planar technologies that evolved from the materials and processes of classical microfabrication to those of large-area electronics on rigid substrates, and then they are branched out to a diverse range of fabrication techniques. Functionalities never seen before can be realised by combining electrical with mechanical properties drawn from all conceivable classes of materials: liquid, gel, solid, organic and inorganic, insulator, semiconductor, and metal.

A Diversity of Materials

The Young¡¯s modulus E of the materials for stretchable electronics spans a range of at least 12 orders of magnitude. In general, both internal energy and entropy change upon mechanical deformation of materials. Materials used in stretchable electronics are comprised of extremes, from entropy elastic elastomers to energy elastic solids. Stretchable electronics combine liquids, soft and elastic materials, and hard and brittle materials. Gels compare to brain tissue (200 Pa) and muscles (1000¨C3000 Pa); the typical polydimethylsiloxane (PDMS) elastomer, electroactive elastomers, and other rubbery materials range from 1 to 100 Mpa; and polymers and organic semiconductors have moduli in the Gpa range. Elastomers with a low Young¡¯s modulus E in the range of 1¨C100 Mpa have a bulk modulus K on the order of 1¨C10 Gpa, which is comparable to those of typical polymers. Such highly elastic materials are easily stretched by mechanical forces, while their volume stays constant. In contrast, solids typically have high Young¡¯s moduli, and they cannot be stretched reversibly by much, or else they will deform plastically or break. Gold (70 Gpa), silicon (150 Gpa), one of the hardest solids, diamond-like carbon (DLC, 500 Gpa), and one of the strongest materials, graphene (1000 Gpa along its plane), complete the spectrum of materials. The regimes of elastic, plastic, and brittle materials also are evident. Combining elastic substrates and interconnects with plastic or brittle device materials sets up tremendous mechanical contrasts, with amazing consequences. A hard solid, when bonded to a polymer, can become surprisingly deformable.

The Architecture of Stretchable Electronics

Stretchable electronics are designed for uniaxial, biaxial, or radial stretching by tens of percent. But optoelectronic devices and circuits contain brittle materials that will break if stretched by more than a few tenths of a percent. For their protection, the circuits are placed on platforms, or islands. Most islands are rigid in-plane and are flexible out-of-plane because they are thinner than ¡«1/1000 of the typical bending radius (e.g., 1 ¦Ì m versus 1 mm). The substrates exposed between the rigid islands, together with the conductors that run across it, accommodate the elastic strain. The islands are made of one of the circuit materials of separate thin-film materials, or of stiff plastic foil. In a macroscopic approach to building stretchable systems, conventional rigid printed circuit board ¡°islands¡± are connected with stretchable Copper conductors in a design similar to that of rigid-flex circuits, where circuit boards are connected by flat cables. When the whole system is stretched, the substrate, devices, interconnects, and protective encapsulating layers interact.

Stretchable Substrates and Encapsulation

We can make substrates stretchable in three ways:

(1) by stretching an elastomeric membrane,

(2) through a net made of an elastomer or of a polymer, or

(3) by stretching the polymer foil of a flex circuit flat.

Polydimethylsiloxane (PDMS) is the elastomer of choice because it can be made with widely accessible and available laboratory techniques, is also commercially available, and only swells in but is not attacked by most process chemicals (although it is etched by KOH and HF). Acrylics also are used as passive substrates similar to PDMS and as electroactive materials. Nets are punched out of polyimide or molded from PDMS. The well-established flex circuits are the equivalent of printed wiring boards but made on plastic foil. The entire flex circuit can be folded and unfolded. Such flex circuits need only bendable instead of elastically stretchable encapsulation. For stretchable organic electronics, some of which are sensitive to atmospheric corrosion, that is an important advantage because no highly hermetic and stretchable permeation barrier has been invented yet.

In stretchable electronics, devices are subject to alternating tensile and compressive strain. This is illustrated with the following example of a membrane that is stretched uniaxially and then relaxed again. Extending a long membrane stripe of length L by pulling out its opposite ends to a length L (1 + ¥ε), where ¥ε is the strain, narrows the stripe from width W to W(1 . ¥ε / 2), and thins it from thickness H to H (1 . ¥ε / 2). Therefore, the device islands and interconnects on the membrane experience lateral compression. When the substrate is relaxed, it experiences lateral tension. The surface area of the substrate increases to L W (1 + ¥ε / 2) during uniaxial stretching to strain ¥ε . Biaxial or radial stretching 44.46 to a linear strain of ¥å raises the surface area to ¡- LW (1 + 2 ¥ε). Hence, for a given value of linear strain ¥ε , biaxially or radially stretching the substrate puts the device islands and interconnects under four times the stress that is developed in uniaxial stretching. In experiments, membranes can be stretched uniaxially to ¥ε = 0.5 and more without breaking devices or interconnects, while going above linear ¥ε . 0.15 in biaxial or radial stretching is a challenge.

For mechanical and chemical protection, stretchable electronics are encapsulated with the same or a similar material and with the same processes, as employed for the substrate. The encapsulant is dispensed as a liquid and is then cured or laminated on. Stretchable biointerfaces on PDMS substrates are encapsulated with photopatternable silicone or with PDMS. Freestanding stretchable conductors are first encased in polyimide and then in PDMS for better mechanical matching with the metal conductor. Contact holes through the encapsulation are opened by plasma etching, liftoff, or lithography of a photopatternable elastomer. Contacts at the edges of the stretchable membrane are opened with procedures similar to liftoff.

Stretchable Interconnects

Stretchable interconnects combine stretchability with high electrical conductance. While thin-film or thinned integrated circuits may need thin-film interconnects, hybrid circuits that include conventional devices can make use of relatively thick printed polymer filled with conducting particles or of copper laminate sheet. Stretchable conductors rely on configurations: waves; meanders; helices of metal on/in/around a stretchable substrate; 54 spirals of silicon; percolation through conducting particles embedded in a stretchable matrix or substrate; nets with meshes on the micrometre, multi-micrometre, or millimetre scales; and liquid metal confined within a microfluidic circuit made of PDMS. These conductors fall into two groups of stretchability up to electrical failure. When placed on or embedded in an elastomer such as PDMS or a gel, some conductors inherently are stretchable to large strains. This is the case with ionimplanted PDMS and carbon nanotube-filled Gel, as well as stretchable gold nets on the micrometre or millimetre scales. Such inherently stretchable conductors need no special preparation for uniaxial or biaxial stretching. A second group of conductors relies on waves that form after a stiff material¡ªa surface coating, conductor, device, or circuit¡ª has been applied to the pre-stretched elastomer. Upon relaxation from pre-stretch, the stiff material buckles to waves as it experiences compression by the contracting substrate. Stretching and relaxing this substrate uniaxially, in the pre-stretch direction, reversibly stretches and relaxes the surface wave.

The shear force at the interface between substrate and device material must not exceed the strength of adhesion, or else the device will peel off when it is compressed during relaxation. When the shear force reaches through the thickness of the substrate, the whole structure will coil up. The highly perfect surface waves that can be made on PDMS lend themselves to tunable diffraction gratings 63 and wavelength-tunable optically pumped lasers.

Stretching pre-formed waves, meanders, and spirals develops smaller strains in the conductor than stretching inherently stretchable materials. 66 When a complete flexible circuit net of silicon islands and interconnects is transfer printed on a chemically patterned, pre-stretched substrate, the silicon links that are coated with interconnect metal arch up to bridges. As the substrate is stretched, these connecting bridges are pulled down toward the substrate. Such up-and-down moving bridges avoid the Poisson compression and expansion of interconnects that are integrated with the elastomer, which may open cracks along the stretching direction of interconnects. One advantage of the surface-integrated conductors is their ease of encapsulation. Preventing the substrate from narrowing during pre-stretch produces fully integrated conductors that are free of cracks. Uniaxial pre-stretching makes the conductor uniaxially stretchable, and biaxial or radial pre-stretching or heating renders it biaxially stretchable. Meanders and spirals can be stretched in only one direction. The deformation of nets on both the micrometre and millimetre scales resembles that of linked meanders, with the links enabling stretching in any direction. Meshes of metallic nets on the micrometre scale have been observed to lengthen and shrink during their stretching and relaxation.

The two important performance parameters for stretchable interconnects are high electrical conductance and large critical strain at which conduction is lost. During stretching, the electrical connection is interrupted when particles disconnect from each other, waves, meanders, spirals, or nets break, or liquid escapes. The critical strain depends on conductor technology and whether the stretching is uniaxial or biaxial. Mechanical prestretching produces structures than can be stretched up to ¡« 100% uniaxially and up to ¡« 15% biaxially. For electrodes on electroactive polymers, a low Young¡¯s modulus, or more precisely, the elastic modulus in the case of a nonlinear elastomer, becomes an equally important third parameter: during their cycles between electrical and mechanical energy, the electrodes must store as little elastic energy as possible, as this energy is robbed from useful conversion. Corrugated metal electrodes meet this requirement.

Devices and Circuits

Broadly speaking, components, devices, and circuits are placed on stretchable substrates by integration or by lamination. For direct integration with the substrate, component materials are deposited and patterned layer-by-layer on the elastomer. This process is similar to the fabrication of conventional wafer-based or thin-film integrated circuits. Patterns are formed either subtractively, that is deposition followed by selective removal, or additively, by the direct printing of patterns. In lamination, components or complete circuits are adhered to the elastomer with adhesives or by transfer printing. The components are pre-fabricated elsewhere on flexible plastic substrates, thin flexible wafers, or rigid islands.

Direct deposition and subtractive processing at present are seen as the pathway to wallpaper-size stretchable systems. Because of their success in flat-panel display fabrication. Highspeed additive printing of optoelectronic circuits eventually may displace these batch fabrication processes. As a substrate, PDMS is quite different from wafer silicon or plate glass. Photolithography and electronic inks employ solvents that make PDMS swell, which in turn may delaminate device films. Converting these processes to alternative solvents takes time. Given the high coefficient of thermal expansion of PDMS, ¡«300 ppm/¡γC, its dimensions may vary noticeably during the temperature cycles associated with processing. Lamination, which has the advantage of minimal on-elastomer processing, produces hybrid circuits that are similar to conventional printed wiring boards and flex circuits. The comparative ease of lamination and hybrid assembly has enabled impressive demonstrations of quite complex stretchable optoelectronic system concepts. Eventually, stretchable systems will combine large integrated arrays of simple devices with nodes of nanoscale integrated circuits in complementary metal-oxide-semiconductor (CMOS) technology to provide large-area, high-resolution, high-performance sensor skins. In this architecture, human-size tasks, such as high-resolution multipoint sensing, will be assumed by electronics directly integrated on the substrate, and tasks that demand high power and high speed computation and communication will be executed by laminated CMOS circuits. Lamination of another kind is sought for biological applications, where the adhesion of healthy cells or of a tissue culture to the stretchable device is a basic measure of the quality of the electronics/ biointerface.

Stretching substrates with rigid circuit islands sets up non-uniform mechanical strain: small strain on and immediately beneath the island, larger strain on the surface between the islands than at the bottom of the elastomer, and largest strain in the elastomer at the periphery of the island. The edge of the island slipped back when the foil was expanded. This slip can occur only if the periphery of the island simultaneously debonds from the substrate. At such transitions between soft and hard materials, surface-integrated interconnects experience their biggest strain and must be designed and fabricated to accommodate it. This situation has been analyzed for plastically deformable substrates and is under study for elastomers. Excessive strain is distributed across soft/hard interfaces by grading mechanical strength. The edges of device islands are feathered out, the elastomer below the island is hardened chemically or by insertion of rigid plates, or the interconnect is made to vary from wide and stiff on the island to narrow and compliant on the elastomer.

Research on stretchable systems has been dominated by the study of fundamental mechanics, development of fabrication technologies, and demonstrations of the functions they enable. For industrialization, a full toolset of process sheets, models, and simulation techniques also will be needed. Excellent initial steps have been taken in mechanical theory and simulation. Many other steps that lead from material to system, including the architecture of electronic surfaces of variable topography, are waiting for exploration by engaged researchers.

A Curvy, Stretchy Future for Electronics

Electronics of the future will be soft and rubbery. Devices based on this new technology will be stretchable, twistable, and deformable into curvilinear shapes, thereby enabling applications that would be impossible to achieve by using the hard, rigid electronics of today. Examples range from ultrathin, conformable health-monitoring tapes that seamlessly mount on the skin, "electronic skin," to advanced imaging devices that use hemispherical detector layouts, "electronic eyeball cameras." In these and other systems, mechanical design will be as important as circuit design.

Historically, progress in electronics has been dominated by a development path first noted by Gordon Moore in 1965: the number of transistors in silicon microprocessors tends to double every 18–24 months, mainly through reductions in the sizes of the transistors. This trend, known as Moore's Law, is now so well established that it is self-fulfilling, via roadmaps that guide the efforts of the entire industry. Modern processors incorporate more than a billion transistors each with critical dimensions of tens of nanometres. Continued downscaling represents an important future for electronics, but not the only one. In the early 1990s, a completely different class of electronics emerged to meet a need that was impossible to address with silicon wafer technology: active matrix circuits for switching pixels in liquid crystal displays. Progress in this kind of electronics is most definitely not dictated by Moore's Law. Here, the primary scaling metric is overall size: bigger is better, because large area coverage means large video screens. The total number of transistors, beyond a few million, and their sizes are both comparatively unimportant. The most advanced fabrication facilities produce circuits that cover areas of nearly 10 m^2, with total outputs that exceed a thousand acres of electronics per year. This "macroelectronics" technology began its growth nearly 40 years after the birth of modern microelectronics, yet it captures a rapidly growing market that now represents a sizable fraction of the entire broader industry.

Many believe that the "next big thing" in electronics is with systems that avoid the rigid, brittle, and planar nature of micro/macroelectronics, to enable new modes of use. Research in this direction began ≈15 years ago, initially driven by interest in flexible, paperlike displays. More recently, the scope has expanded to include more compelling, and more technically challenging, opportunities in biomedical devices that intimately integrate with the human body and in product designs that exploit curvilinear, ergonomic, or biologically inspired layouts. One approach to such technology relies on the development of new electronic materials, e.g., organic semiconductors that can flex and composite elastomer conductors that can stretch. Different strategies use optimized structural configurations of established inorganic materials for stretchable interconnects and/or active devices. The concepts that enable stretchy properties from these brittle materials are simple. For example, ultrathin sheets of silicon are flexible, for the same reason that any material in thin film form is flexible: bending strains are inversely proportional to thickness. Such sheets, formed in buckled "wavy" shapes and integrated with rubber substrates, yield composite systems that provide fully elastic responses to applied strains, with an underlying physics similar to that of an accordion bellows. These and related ideas provide routes to high-performance electronics with the mechanical properties of a rubber band. In parallel with fundamental research on the materials and mechanics, advanced development work is increasing in areas of biomedicine such as intracardial and neural monitoring. As applied to digital imaging, this technology enables hemispherical electronic eyeball cameras that provide enhanced operating characteristics compared with devices with conventional planar layouts.

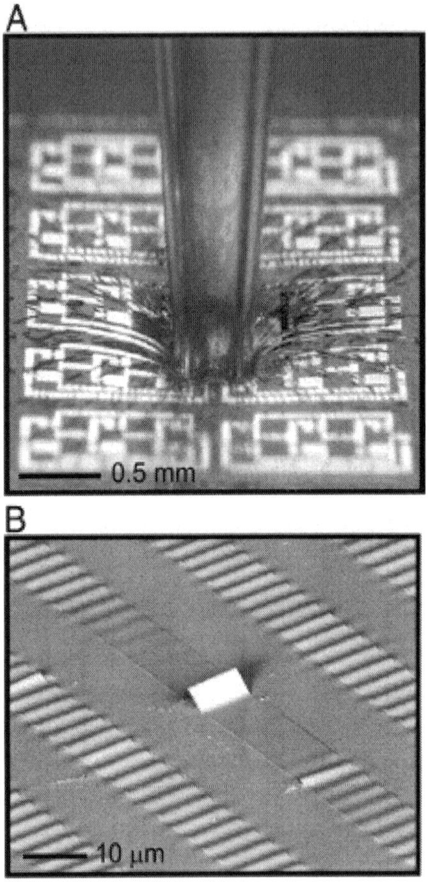

Figure: *View larger version:*

Buckled ultrathin silicon circuit sheets bonded to elastomeric substrates represent a form of electronics that responds elastically to applied strains. (*A*) Optical image of a silicon CMOS (complementary metal–oxide semiconductor) integrated circuit of this type, pressed downward in the central region with a glass rod. The wavelengths and amplitudes of the buckled structures, visible here as colour and contrast variations, change to accommodate the deformation. This physics provides a useful degree of circuit-level stretchability that avoids any significant strains in the electronic materials themselves. (*B*) Scanning electron micrograph of buckled nanoribbons of single crystalline silicon bonded to an elastomer substrate. This composite system provides an elastic response to applied strains, with a physics that is conceptually similar to that of an accordion bellows. The ribbon in the centre of this image had a defect that caused it to delaminate locally from the substrate to form an isolated bridge structure, further highlighting certain aspects of the mechanics. This stretchable form of silicon embodies, in a simple system, the key physics associated with the full integrated circuit shown in *A*.

Mechanical design plays a critically important role in this type of stretchable electronics technology. Experimental work involves characterizing the microscopic behaviour of the material structures, the devices, circuits, and systems, where the length scales for the relevant deformations range from millimetres to nanometres Analytical and computational modelling can quantitatively capture the physics of this basic subunit of stretchable integrated circuits. Although such studies of the mechanics are critically important, they require sophisticated capabilities in device fabrication and test. Vella et al. show that the essential physics can be explored by using macroscopic analogs, consisting of plastic sheets on rubber supports, manipulated with clamps and translation stages.

This approach greatly simplifies the experiments, thereby opening up the field to researchers who do not have access to cleanroom fabrication facilities or advanced microscope systems. Vella et al. exploit their setups to study the nonlinear processes that lead to buckled structures. Their mathematical treatment uses established methods of energy minimization, but with a unique linking of the release of total system energy to the interfacial fracture toughness. The result provides a useful way to determine not only the onset but also the evolution of buckling with applied strain. They use macroscopically measurable interfacial toughness to determine microscopic parameters such as buckling and wavelength and amplitude.

One outcome of this analysis is the recognition of a critical film thickness needed to avoid damage or plastic deformation, thereby providing an important design rule for devices. Explicitly accounting for the loading and geometry dependence of the interfacial fracture toughness in the theory will further improve the accuracy of the results.

Microactuator

A microactuator is a microscopic servomechanism that supplies and transmits a measured amount of energy for the operation of another mechanism or system. As a general actuator, following standards have to be met.

- Large travel
- High precision
- Fast switching
- Low power consumption
- Power free force sustainability

For microactuator, there are two in addition

- Microstructurability
- Integrability

Principle of Microactuators

The basic principle can be described as the expression for mechanical work

$$W = \vec{F} \cdot \Delta\vec{r}$$

since an actuator is to manipulate positions and therefore force is needed. For different kind of microactuators, different physical principles are applied.

Classes of Microactuators

- Electrostatic
- Electromagnetic
- Piezoelectric
- Fluid
- Thermal

Elastomer Based Pressure and Strain Sensors

An elastomer is a polymer with viscoelasticity (colloquially "elasticity"), generally having low Young's modulus and high failure strain compared with other materials. The term, which is derived from *elastic polymer*, is often used interchangeably with the term rubber, although the latter is preferred when referring to vulcanisates.

Each of the monomers which link to form the polymer is usually made of carbon, hydrogen, oxygen and/or silicon. Elastomers are amorphous polymers existing above their glass transition temperature, so that considerable segmental motion is possible. At ambient temperatures, rubbers are thus relatively soft (E~3MPa) and deformable. Their primary uses are for seals, adhesives and molded flexible parts. Application areas for different types of rubber are manifold and cover segments as diverse as tires, shoe soles as well as dampening and insulating elements. The importance rubbers have can be judged from the fact that global revenues are forecast to rise to US$56 billion in 2020.

Elastomers are usually thermosets (requiring vulcanization) but may also be thermoplastic. The long polymer chains cross-link during curing, i.e., vulcanizing. The molecular structure of elastomers can be imagined as a 'spaghetti and meatball' structure, with the meatballs signifying cross-links. The elasticity is derived from the ability of the long chains to reconfigure themselves to distribute an applied stress.

The covalent cross-linkages ensure that the elastomer will return to its original configuration when the stress is removed. As a result of this extreme flexibility, elastomers can reversibly extend from 5-700%, depending on the specific material. Without the cross-linkages or with short, uneasily reconfigured chains, the applied stress would result in a permanent deformation.

Temperature effects are also present in the demonstrated elasticity of a polymer. Elastomers that have cooled to a glassy or crystalline phase will have less mobile chains, and consequentially less elasticity, than those manipulated at temperatures higher than the glass transition temperature of the polymer.

It is also possible for a polymer to exhibit elasticity that is not due to covalent cross-links, but instead for thermodynamic reasons.

Examples of Elastomers

Unsaturated rubbers that can be cured by sulfur vulcanization:

- Natural polyisoprene: cis-1,4-polyisoprene natural rubber (NR) and trans-1,4-polyisoprene gutta-percha
- Synthetic polyisoprene (IR for Isoprene Rubber)
- Polybutadiene (BR for Butadiene Rubber)
- Chloropene rubber (CR), polychloroprene, Neoprene, Baypren etc.
- Butyl rubber (copolymer of isobutylene and isoprene, IIR)
    - Halogenated butyl rubbers (chloro butyl rubber: CIIR; bromo butyl rubber: BIIR)
- Styrene-butadiene Rubber (copolymer of styrene and butadiene, SBR)
- Nitrile rubber (copolymer of butadiene and acrylonitrile, NBR), also called Buna N rubbers
    - Hydrogenated Nitrile Rubbers (HNBR) Therban and Zetpol

(Unsaturated rubbers can also be cured by non-sulfur vulcanization if desired).

Saturated rubbers that cannot be cured by sulfur vulcanization:

- EPM (ethylene propylene rubber, a copolymer of ethylene and propylene) and EPDM rubber (ethylene propylene diene rubber, a terpolymer of ethylene, propylene and a diene-component)
- Epichlorohydrin rubber (ECO)
- Polyacrylic rubber (ACM, ABR)
- Silicone rubber (SI, Q, VMQ)
- Fluorosilicone Rubber (FVMQ)
- Fluoroelastomers (FKM, and FEPM) Viton, Tecnoflon, Fluorel, Aflas and Dai-El
- Perfluoroelastomers (FFKM) Tecnoflon PFR, Kalrez, Chemraz, Perlast
- Polyether block amides (PEBA)
- Chlorosulfonated polyethylene (CSM), (Hypalon)
- Ethylene-vinyl acetate (EVA)

"The definitions are not authentic as the Rubber which is classified in World Customs Organisation Books in Chapter 40, where as the above definitions stating all rubber and different polymers in same chapter which is classified in Chapter 39 of the World Custom Organisation's Harmonised Commodity for Description and coding system.

One should go through all differentiation while editing between Plastics and articles thereof and Rubber and articles thereof."

Various other types of elastomers:

- Thermoplastic elastomers (TPE)
- The proteins resilin and elastin
- Polysulfide rubber
- Elastolefin, elastic fibre used in fabric production

Conformable Active Devices

Organic field-effect transistors and their integratedcircuits have attracted considerable attention because of attributes that complement silicon-based large-scale integrations, devices that are high-performance but expensive.

Organic transistors can be manufactured on plastic films at low (ambient) temperatures; therefore, they are mechanically flexible and potentially inexpensive to manufacture. Two major applications drive recent studies of organic transistors.

The first includes flexible displays such as paper-like displays or e-paper, in which electric inks or other media are driven by organic transistor active matrixes. The second comprises radio frequency identification tags. The printable features of organic transistors should facilitate the implementation of radio frequency identification on packages.

Although the mobility of organic semicon-ductors is known to be about two or three orders of magnitude less than that of poly- and single-crystalline silicon, the slower speed is tolerable for most applications of large-area sensors.

For E-skin in particular, the integration of pressure sensors and organic peripheral electronics allows the manufacturer to avoid the drawbacks of organic transistors while taking advantage of their many benefits, such as mechanical flexibility, large area, low cost, and relative ease of fabrication.

Although the development of flexible pressure sensor films was an important step in endowing robots with skin sensitivity, E-skins are expected to add at least two more functionalities, namely, thermal sensing and conformability. Without conformability, it is impossible to apply E-skin to three-dimensional surfaces.

Stretchable artificial skins for humans are now commercially available, but they lack electric functionality. Indeed, various stretchable materials, such as rubber, are used in daily activities, but they have poor electrical conductivity. One of the biggest challenges in the field of electronics, particularly for E-skin applications, is the manufacture of stretchable active electronic elements and interconnects.

Although the base films that we presently use are of polyimide and poly(ethylenenaphthalate), which are stiff and not inherently stretchable (in a rubber-like sense), our solution includes struts of network structures that twist with the application of tension. Because of this three-dimensional strut deformation, the whole network structure functions electrically with a unidirectional extension of 25%.

5

Photovoltaics

Photovoltaics (PV) is a method of generating electrical power by converting solar radiation into direct current electricity using semiconductors that exhibit the photovoltaic effect. Photovoltaic power generation employs solar panels composed of a number of solar cells containing a photovoltaic material. Mainstream materials presently used for photovoltaics includc monocrystalline silicon, polycrystalline silicon, amorphous silicon, cadmium telluride, and copper indium gallium selenide/sulfide. Due to the increased demand for renewable energy sources, the manufacturing of solar cells and photovoltaic arrays has advanced considerably in recent years.

Solar photovoltaics is a sustainable energy source. In 2013, its fast-growing capacity increased by 36% to a running total of 136 GW, worldwide. This is sufficient to generate 160 TWh/year or about 0.85% of the electricity demand on the planet. China and Japan are now the fastest growing markets, while Germany remains the world's largest producer, contributing almost 6% to its national electricity demands.

Solar photovoltaics is now, after hydro and wind power, the third most important renewable energy source in terms of globally installed capacity. More than 100 countries use solar PV. Installations may be ground-mounted (and sometimes integrated with farming and grazing) or built into the roof or walls of a building (either building-integrated photovoltaics or simply rooftop).

Driven by advances in technology and increases in manufacturing scale and sophistication, the cost of photovoltaics has declined steadily since the first solar cells were manufactured, and the levelised cost of electricity (LCOE) from

PV is competitive with conventional electricity sources in an expanding list of geographic regions. Net metring and financial incentives, such as preferential feed-in tariffs for solar-generated electricity, have supported solar PV installations in many countries. With current technology, photovoltaics recoup the energy needed to manufacture them in 1.5 (in Southern Europe) to 2.5 years (in Northern Europe).

Solar Cells

Photovoltaics are best known as a method for generating electric power by using solar cells to convert energy from the sun into a flow of electrons. The photovoltaic effect refers to photons of light exciting electrons into a higher state of energy, allowing them to act as charge carriers for an electric current. The photovoltaic effect was first observed by Alexandre-Edmond Becquerel in 1839. The term photovoltaic denotes the unbiased operating mode of a photodiode in which current through the device is entirely due to the transduced light energy. Virtually all photovoltaic devices are some type of photodiode.

Solar cells produce direct current electricity from sun light which can be used to power equipment or to recharge a battery. The first practical application of photovoltaics was to power orbiting satellites and other spacecraft, but today the majority of photovoltaic modules are used for grid connected power generation. In this case an inverter is required to convert the DC to AC. There is a smaller market for off-grid power for remote dwellings, boats, recreational vehicles, electric cars, roadside emergency telephones, remote sensing, and cathodic protection of pipelines.

Photovoltaic power generation employs solar panels composed of a number of solar cells containing a photovoltaic material. Materials presently used for photovoltaics include monocrystalline silicon, polycrystalline silicon, amorphous silicon, cadmium telluride, and copper indium gallium selenide/sulfide. Copper solar cables connect modules (module cable), arrays (array cable), and sub-fields. Because of the growing demand for renewable energy sources, the manufacturing of solar cells and photovoltaic arrays has advanced considerably in recent years.

Cells require protection from the environment and are usually packaged tightly behind a glass sheet. When more power is required than a single cell can deliver, cells are electrically connected together to form photovoltaic modules, or solar panels. A single module is enough to power an emergency telephone, but for a house or a power plant the modules must be arranged in multiples as arrays.

Photovoltaic power capacity is measured as maximum power output under standardized test conditions (STC) in "Wp" (Watts peak). The actual power output at a particular point in time may be less than or greater than this standardized, or "rated," value, depending on geographical location, time of day, weather conditions, and other factors. Solar photovoltaic array capacity factors are typically under 25%, which is lower than many other industrial sources of electricity.

A significant market has emerged in off-grid locations for solar-power-charged storage-battery based solutions. These often provide the only electricity available. The first commercial installation of this kind was in 1966 on Ogami Island in Japan to transition Ogami Lighthouse from gas torch to fully self-sufficient electrical power.

Current Developments

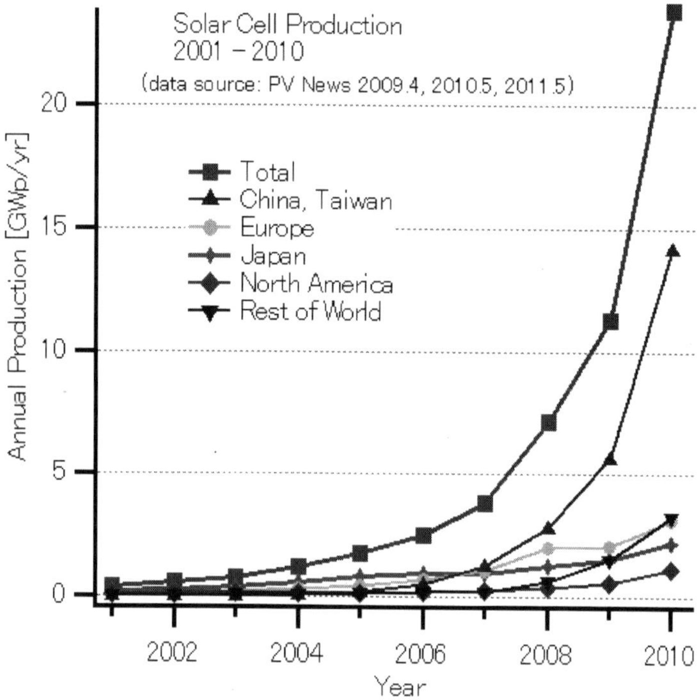

Figure: *Solar cell productions by region*

Photovoltaic Powerworldwide GWp	
2000	1
2005	5
2010	40
2011	70
2012	99
2013	136
Year end capacities	

For best performance, terrestrial PV systems aim to maximize the time they face the sun. Solar trackers achieve this by moving PV panels to follow the sun. The increase can be by as much as 20% in winter and by as much as 50% in summer. Static mounted systems can be optimized by analysis of the sun path. Panels are often set to latitude tilt, an angle equal to the latitude, but performance can be improved by adjusting the angle for summer or winter. Generally, as with other semiconductor devices, temperatures above room temperature reduce the performance of photovoltaics.

A number of solar panels may also be mounted vertically above each other in a tower, if the zenith distance of the Sun is greater than zero, and the tower can be turned horizontally as a whole and each panels additionally around a horizontal axis. In such a tower the panels can follow the Sun exactly. Such a device may be described as a ladder mounted on a turnable disk. Each step of that ladder is the middle axis of a rectangular solar panel. In case the zenith distance of the Sun reaches zero, the "ladder" may be rotated to the north or the south to avoid a solar panel producing a shadow on a lower solar panel. Instead of an exactly vertical tower one can choose a tower with an axis directed to the polar star, meaning that it is parallel to the rotation axis of the Earth. In this case the angle between the axis and the Sun is always larger than 66 degrees. During a day it is only necessary to turn the panels around this axis to follow the Sun.

Solar photovoltaics is growing rapidly, albeit from a small base, to a total global capacity of 136 Gigawatts (GW) at the end of 2013. The total power output of the world's PV capacity in a calendar year is equal to some 160 billion kWh of electricity. This is sufficient to cover the annual power supply needs of 40 million households in the world, and represents 0.85% of worldwide electricity demand. More than 100 countries use solar PV. World solar PV capacity (grid-connected) was 7.6 GW in 2007, 16 GW in 2008, 23 GW in 2009, and 40 GW in 2010. Installations may be ground-mounted (and sometimes integrated with farming and grazing) or built into the roof or walls of a building (building-integrated photovoltaics). Photovoltaics is now, after hydro and wind power, the third most important renewable energy source in terms of globally installed capacity.

The 2011 European Photovoltaic Industry Association (EPIA) report predicted that, "[In 2011] Europe once again was the global leader in PV market growth, with 75% of all newly connected capacity and about 75% of global installed capacity. But non-European markets are showing signs that they may soon shift this balance in their favour". With proper policy support, balanced market development, and continued industry innovation, photovoltaic (PV) can continue its remarkable growth rate over the short-, medium- and long-term, and even beyond.

The EPIA/Greenpeace Solar Generation Paradigm Shift Scenario (formerly called Advanced Scenario) from 2010 shows that by the year 2030, 1,845 GW of PV systems could be generating approximately 2,646 TWh/year of electricity around the world. Combined with energy use efficiency improvements, this would represent the electricity needs of more than 9% of the world's population. By 2050, over 20% of all electricity could be provided by photovoltaics.

However, the EPIA prediction may be pessimistic since official agencies keep underestimating the growth rate of renewables. A report based on the 2012 BP Statistical Review shows an exponential growth in global solar generation from 2001 to end 2011, with an approximate doubling of generation every two years. This raises the possibility that solar power could reach 10% of total global power generation by the end of this decade. To accomplish this gain in primary energy share, solar will need to advance from the 55.7 TWh generated in 2011 to approximately 2200 TWh. At current exponential growth rates, those levels could be achieved as early as 2018 rather than around 2030 as suggested by the EPIA. Solar would provide 100 percent of the current world energy needs by 2027 if the biannual doubling of generation continues.

The San Jose-based company Sunpower produces cells that have an energy conversion ratio of 19.5%, well above the market average of 12–18%. The most efficient solar cell so far is a multi-junction concentrator solar cell with an efficiency of 43.5% produced by Solar Junction in April 2011. The highest efficiencies achieved without concentration include Sharp Corporation at 35.8% using a proprietary triple-junction manufacturing technology in 2009, and Boeing Spectrolab (40.7% also using a triple-layer design). A March 2010 experimental demonstration of a design by a Caltech group led by Harry Atwater which has an absorption efficiency of 85% in sunlight and 95% at certain wavelengths is claimed to have near perfect quantum efficiency. However, absorption efficiency should not be confused with the sunlight-to-electricity conversion efficiency.

Several companies have begun embedding power optimizers into PV modules called "smart modules". These modules perform maximum power point tracking (MPPT) for each module individually, measure performance data for monitoring, and provide additional safety. Such modules can also compensate for shading effects, wherein a shadow falling across a section of a module causes the electrical output of one or more strings of cells in the module to fall to zero, but not having the output of the entire module fall to zero.

At the end of September 2013, IKEA announced that solar panel packages for houses will be sold at 17 United Kingdom IKEA stores by the end of July 2014. The decision followed a successful pilot project at the Lakeside IKEA store, whereby one photovoltaic (PV) system was sold almost every day. The panels are manufactured by a Chinese company named Hanergy Holding Group Ltd.

Economics

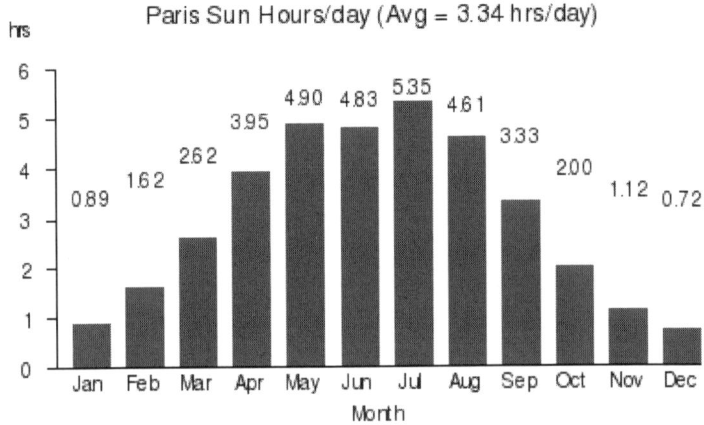

Source: Apricus

Financial incentives for photovoltaics, such as feed-in tariffs, have often been offered to electricity consumers to install and operate solar-electric generating systems. Government has sometimes also offered incentives in order to encourage the PV industry to achieve the economies of scale needed to compete where the cost of PV-generated electricity is above the cost from the existing grid. Such policies are implemented to promote national or territorial energy independence, high tech job creation and reduction of carbon dioxide emissions which cause global warming. Due to economies of scale solar panels get less costly as people use and buy more—as manufacturers increase production to meet demand, the cost and price is expected to drop in the years to come.

Solar cell efficiencies vary from 6% for amorphous silicon-based solar cells to 44.0% with multiple-junction concentrated photovoltaics. Solar cell energy conversion efficiencies for commercially available photovoltaics are around 14–22%. Concentrated photovoltaics (CPV) may reduce cost by concentrating up to 1,000 suns (through magnifying lens) onto a smaller sized photovoltaic cell. However, such concentrated solar power requires sophisticated heat sink designs, otherwise the photovoltaic cell overheats, which reduces its efficiency and life. To further exacerbate the concentrated cooling design, the heat sink must be passive, otherwise the power required for active cooling would reduce the overall efficiency and economy.

Crystalline silicon solar cell prices have fallen from $76.67/Watt in 1977 to an estimated $0.74/Watt in 2013. This is seen as evidence supporting Swanson's law, an observation similar to the famous Moore's Law that states that solar cell prices fall 20% for every doubling of industry capacity.

As of 2011, the price of PV modules per MW has fallen by 60% since the summer of 2008, according to Bloomberg New Energy Finance estimates, putting solar power for the first time on a competitive footing with the retail price of electricity in a number of sunny countries; an alternative and consistent price decline figure of 75% from 2007 to 2012 has also been published, though it is unclear whether these figures are specific to the United States or generally global. The levelised cost of electricity (LCOE) from PV is competitive with conventional electricity sources in an expanding list of geographic regions, particularly when the time of generation is included, as electricity is worth more during the day than at night. There has been fierce competition in the supply chain, and further improvements in the levelised cost of energy for solar lie ahead, posing a growing threat to the dominance of fossil fuel generation sources in the next few years. As time progresses, renewable energy technologies generally get cheaper, while fossil fuels generally get more expensive:

The less solar power costs, the more favourably it compares to conventional power, and the more attractive it becomes to utilities and energy users around the globe. Utility-scale solar power can now be delivered in California at prices well below $100/MWh ($0.10/kWh) less than most other peak generators, even those running on low-cost natural gas. Lower solar module costs also stimulate demand from consumer markets where the cost of solar compares very favourably to retail electric rates.

As of 2011, the cost of PV has fallen well below that of nuclear power and is set to fall further. The average retail price of solar cells as monitored by the Solarbuzz group fell from $3.50/watt to $2.43/watt over the course of 2011.

For large-scale installations, prices below $1.00/watt were achieved. A module price of 0.60 Euro/watt ($0.78/watt) was published for a large scale 5-year deal in April 2012.

By the end of 2012, the "best in class" module price had dropped to $0.50/watt, and was expected to drop to $0.36/watt by 2017.

In many locations, PV has reached grid parity, which is usually defined as PV production costs at or below retail electricity prices (though often still above the power station prices for coal or gas-fired generation without their distribution and other costs). However, in many countries there is still a need for more access to capital to develop PV projects. To solve this problem securitization has been proposed and used to accelerate development of solar photovoltaic projects. For example, SolarCity offered, the first U.S. asset backed security in the solar industry in 2013.

Photovoltaic power is also generated during a time of day that is close to peak demand (precedes it) in electricity systems with high use of air conditioning. More generally, it is now evident that, given a carbon price of $50/ton, which would raise the price of coal-fired power by 5c/kWh, solar PV will be cost-competitive in most locations. The declining price of PV has been reflected in rapidly growing installations, totaling about 23 GW in 2011. Although some consolidation is likely in 2012, due to support cuts in the large markets of Germany and Italy, strong growth seems likely to continue for the rest of the decade. Already, by one estimate, total investment in renewables for 2011 exceeded investment in carbon-based electricity generation.

In the case of self consumption payback time is calculated based on how much electricity is not brought from the grid. Additionally, using PV solar power to charge DC batteries, as used in Plug-in Hybrid Electric Vehicles and Electric Vehicles, leads to greater efficiencies. Traditionally, DC generated electricity from solar PV must be converted to AC for buildings, at an average 10% loss during the conversion. An additional efficiency loss occurs in the transition back to DC for battery driven devices and vehicles, and using various interest rates and energy price changes were calculated to find present values that range from $2,057.13 to $8,213.64 (analysis from 2009).

For example in Germany with electricity prices of 0.25 euro/KWh and Insolation of 900 KWh/KW one KWp will save 225 euro per year and with installation cost of 1700 euro/KWp means that the system will pay back in less than 7 years.

Organic Photovoltaics

An organic solar cell or plastic solar cell is a type of polymer solar cell that uses organic electronics, a branch of electronics that deals with conductive organic polymers or small organic molecules, for light absorption and charge transport to produce electricity from sunlight by the photovoltaic effect.

The plastic used in organic solar cells has low production costs in high volumes. Combined with the flexibility of organic molecules, organic solar cells are potentially cost-effective for photovoltaic applications. Molecular engineering (e.g. changing the length and functional group of polymers) can change the energy gap, which allows chemical change in these materials. The optical absorption coefficient of organic molecules is high, so a large amount of light can be absorbed with a small amount of materials. The main disadvantages associated with organic photovoltaic cells are low efficiency, low stability and low strength compared to inorganic photovoltaic cells.

Organic Photovoltaic Materials

A photovoltaic cell is a specialised semiconductor diode that converts visible light into direct current (DC) electricity. Some photovoltaic cells can also convert infrared (IR) or ultraviolet (UV) radiation into DC. A common characteristic of both the small molecules and polymers used in photovoltaics is that they all have large conjugated systems. A conjugated system is formed where carbon atoms covalently bond with alternating single and double bonds; in other words these are chemical reactions of hydrocarbons. These hydrocarbons' electrons pz orbitals delocalize and form a delocalized bonding π orbital with a π* antibonding orbital. The delocalized π orbital is the highest occupied molecular orbital (HOMO), and the π* orbital is the lowest unoccupied molecular orbital (LUMO). The separation between HOMO and LUMO is considered the band gap of organic electronic materials. The band gap is typically in the range of 1–4 eV.

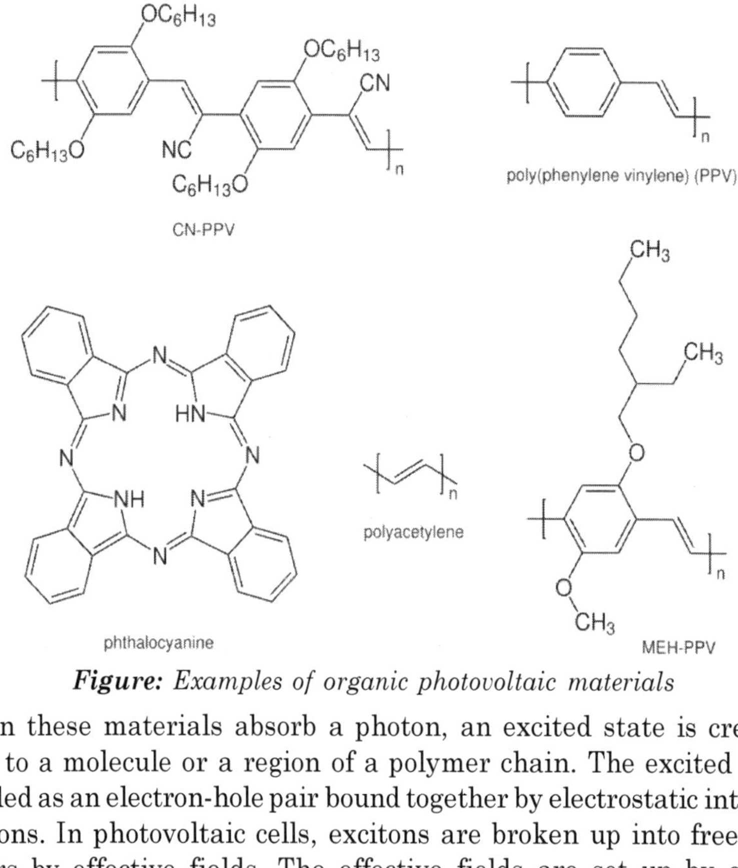

Figure: *Examples of organic photovoltaic materials*

When these materials absorb a photon, an excited state is created and confined to a molecule or a region of a polymer chain. The excited state can be regarded as an electron-hole pair bound together by electrostatic interactions, i.e. excitons. In photovoltaic cells, excitons are broken up into free electron-hole pairs by effective fields. The effective fields are set up by creating a heterojunction between two dissimilar materials. Effective fields break up excitons by causing the electron to fall from the conduction band of the absorber to the conduction band of the acceptor molecule. It is necessary that the acceptor material has a conduction band edge that is lower than that of the absorber material.

Types of Junctions for OPVC

Single Layer Organic Photovoltaic Cell

Single layer organic photovoltaic cells are the simplest of the various forms of organic photovoltaic cells. These cells are made by sandwiching a layer of organic electronic materials between two metallic conductors, typically a layer of indium tin oxide (ITO) with high work function and a layer of low work function metal such as Al, Mg or Ca.

The difference of work function between the two conductors sets up an electric field in the organic layer. When the organic layer absorbs light, electrons will be excited to the LUMO and leave holes in the HOMO, thereby forming excitons. The potential created by the different work functions helps to separate the exciton pairs, pulling electrons to the positive electrode (an electrical conductor used to make contact with a non-metallic part of a circuit) and holes to the negative electrode. The current and voltage resulting from this process can be used to do work. Using electric fields is not the best way to break up excitons: heterojunction based cells which rely on effective fields are more effective.

Examples

Cells with phthalocyanine as an organic layer were investigated at the early stage. As early as 1958, Kearns et al. reported the photovoltaic effect or the creation of voltage of a cell based on magnesium phthalocyanine, a macrocyclic compound having an alternating nitrogen atom-carbon atom ring structure (MgPh), which had a photovoltage of 200 mV. Ghosh et al. investigated the Al/MgPh/Ag cell, and obtained photovoltaic efficiency of 0.01% under illumination at 690 nm.

Conjugated polymers were also used in this type of photovoltaic cell. Weinberger et al. used polyacetylene as the organic layer, Al and graphite as electrodes to fabricate a cell, which had an open circuit voltage of 0.3 V and a charge collection efficiency of 0.3%. Glenis et al. reported an Al/poly(3-nethyl-thiophene)/Pt cell had an external quantum yield of 0.17%, an open circuit voltage of 0.4 V and a fill factor of 0.3. Karg et al. fabricated an ITO/PPV/Al cell, showing an open circuit voltage of 1 V and a power conversion efficiency of 0.1% under white-light illumination.

Problems

In practice, single layer organic photovoltaic cells of this type do not work well. They have low quantum efficiencies (<1%) and low power conversion efficiencies (<0.1%). A major problem with them is that the electric field resulting from the difference between the two conductive electrodes is seldom sufficient to break up the photogenerated excitons. Often the electrons recombine with the holes rather than reach the electrode. To deal with this problem, the multilayer organic photovoltaic cells were developed.

Bilayer Organic Photovoltaic Cells

This type of organic photovoltaic cell contains two different layers in between the conductive electrodes. These two layers of materials have differences in electron affinity and ionization energy, therefore electrostatic forces are generated at the interface between the two layers. The materials are chosen properly to make the differences large enough, so these local electric fields are strong, which may break up the excitons much more efficiently than the single layer photovoltaic cells do. The layer with higher electron affinity and ionization potential is the electron acceptor, and the other layer is the electron donor. This structure is also called a planar donor-acceptor heterojunction.

Examples

C_{60} has high electron affinity, making it a good electron acceptor in photovoltaic cells of this type. Sariciftci et al. fabricated a C_{60}/MEH-PPV double layer cell, which had a relatively high fill factor of 0.48 and a power conversion efficiency of 0.04% under monochromatic illumination. For PPV/C_{60} cells, Halls et al. reported a monochromatic external quantum efficiency of 9%, a power conversion efficiency of 1% and a fill factor of 0.48.

Perylene derivatives are a group of organic molecules with high electron affinity and chemical stability. Ching W. Tang deposited a layer of copper phthalcocyanine as electron donor and perylene tetracarboxylic derivative as electron acceptor, fabricating a cell with a fill factor as high as 0.65 and a power conversion efficiency of 1% under simulated AM2 illumination. Halls et al. fabricated a cell with a layer of bis(phenethylimido) perylene over a layer of PPV as the electron donor. This cell had peak external quantum efficiency of 6% and power conversion efficiency of 1% under monochromatic illumination, and a fill factor of up to 0.6.

Problems

The diffusion length of excitons in organic electronic materials is typically on the order of 10 nm. In order for most excitons to diffuse to the interface of layers and break up into carriers, the layer thickness should also be in the same range as the diffusion length. However, a polymer layer typically needs a thickness of at least 100 nm to absorb enough light. At such a large thickness, only a small fraction of the excitons can reach the heterojunction interface. To address this problem, a new type of heterojunction photovoltaic cells is designed, which is the dispersed heterojunction photovoltaic cells.

Bulk Heterojunction Photovoltaic Cells

In this type of photovoltaic cell, the electron donor and acceptor are mixed together, forming a polymer blend. If the length scale of the blend is similar to the exciton diffusion length, most of the excitons generated in either material may reach the interface, where excitons break efficiently. Electrons move to the acceptor domains then were carried through the device and collected by one electrode, and holes were pulled in the opposite direction and collected at the other side.

While most bulk heterojunction cells are of the two component variety, three component cells have also been explored. The third component, a secondary p-type donor polymer, acts to absorb light in a region of the solar spectrum different from that of the other donor. This in theory helps to maximize the amount of light absorbed by the cell.

These ternary cells can operate through one of three distinct mechanisms: charge transfer, energy transfer, or parallel-linkage. In charge transfer, both donors contribute directly to the generation of free charge carriers. Holes pass through only one donor domain before collection at the anode. In energy transfer, only one donor contributes to the production of holes. The second donor acts solely to absorb light, transferring extra photon energy to the first donor material. In parallel linkage, both donors produce excitons independently, which then migrate to their respective donor/acceptor interfaces and dissociate.

Examples

C_{60} and its derivatives are also used as electron acceptor in the dispersed heterojunction photovoltaic cells. Yu et al. fabricated a cell with the blend of MEH-PPV and a methano-functionalized C_{60} derivative as the heterojunction, ITO and Ca as the electrodes. This cell showed a quantum efficiency of 29% and a power conversion efficiency of 2.9% under monochromatic illumination. Later they replaced MEH-PPV with P3HT, which obtained a cell with a quantum yield of 45% under a 10 V reverse bias.

Polymer/polymer blends are also used in dispersed heterojunction photovoltaic cells. Halls et al. used a blend of CN-PPV and MEH-PPV, fabricated a cell with Al and ITO as the electrodes, whose peak monochromatic power conversion efficiency is 1% and fill factor is 0.38.

Dye sensitized photovoltaic cells can also be considered important examples of this type.

Graded Heterojunction Photovoltaic Cells

In this type of photovoltaic cell, the electron donor and acceptor are mixed together, like in the bulk heterojunction, but in such a way that the gradient is gradual. This architecture combines the short electron travel distance in the dispersed heterojunction with the advantage of the charge gradient of the bilayer technology.

Examples

Holmes et al. fabricated a cell with a blend of CuPc and C_{60}. The cell showed a quantum efficiency of 50% and a power conversion efficiency of 2.1% using 100 mW/cm^2 simulated AM1.5G solar illumination for a graded heterojunction.

Current Challenges and Recent Progress

Difficulties associated with organic photovoltaic cells include their low external quantum efficiency (up to 70%) in comparison with inorganic photovoltaic devices; due largely to the large band gap of organic materials. Instabilities against oxidation and reduction, recrystallization and temperature variations can also lead to device degradation and decreased performance over time. This occurs to different extents for devices with different compositions, and is an area into which active research is taking place.

Other important factors include the exciton diffusion length; charge separation and charge collection; and charge transport and mobility, which are affected by the presence of impurities.

Effect of Film Morphology

As described above, dispersed heterojunctions of donor-acceptor organic materials have high quantum efficiencies compared to the planar hetero-junction, because in dispersed heterojunctions it is more likely for an exciton to find an interface within its diffusion length. Film morphology can also have a drastic effect on the quantum efficiency of the device. Rough surfaces and the presence of voids can increase the series resistance and also the chance of short-circuiting. Film morphology and, as a result, quantum efficiency can be improved by annealing of a device after covering it by a ~1000 E thick metal cathode. Metal film on top of the organic film applies stresses on the organic film, which helps to prevent the morphological relaxation in the organic film. This gives more densely packed films and at the same time allows the formation of phase-separated interpenetrating donor-acceptor interface inside the bulk of organic thin film.

Controlled Growth Heterojunction

Charge separation occurs at the donor acceptor interface. Whilst travelling to the electrode, a charge can become trapped and/or recombine in a disordered interpenetrating organic material, resulting in decreased device efficiency. Controlled growth of the heterojunction provides better control over positions of the donor-acceptor materials, resulting in much greater power efficiency (ratio of output power to input power) than that of planar and highly disoriented hetero-junctions. Thus, the choice of suitable processing parameters in order to better control the structure and film morphology is highly desirable.

Progress in Growth Techniques

Mostly organic films for photovoltaic applications are deposited by spin coating and vapor-phase deposition. However each method has certain draw backs, spin coating technique can coat larger surface areas with high speed but the use of solvent for one layer can degrade the already existing polymer layer. Another problem is related with the patterning of the substrate for device as spin-coating results in coating the entire substrate with a single material.

Vacuum Thermal Evaporation

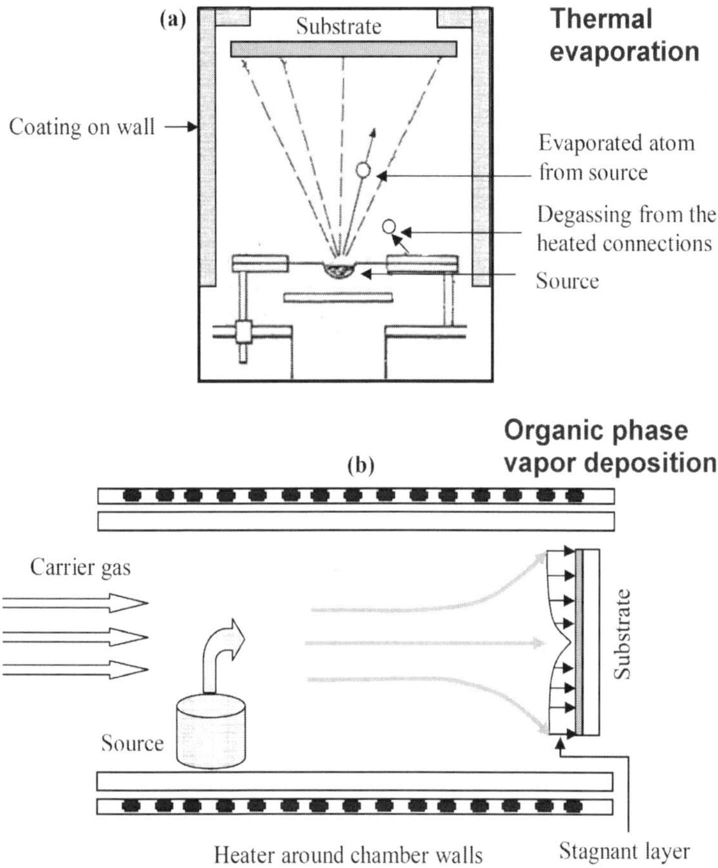

Figure: *Vacuum thermal evaporation (a) and organic phase vapor deposition (b)*

Another deposition technique is vacuum thermal evaporation (VTE) which involves the heating of an organic material in vacuum. The substrate is placed several centimetres away from the source so that evaporated material may be directly deposited onto the substrate. This method is useful for depositing many layers of different materials without chemical interaction between different layers. However, there are sometimes problems with film-thickness uniformity and uniform doping over large-area substrates. In addition, the materials that deposit on the wall of the chamber can contaminate later depositions. This “line of sight” technique also can create holes in the film due to shadowing, which causes an increase in the device series-resistance and short circuit.

Organic Vapor Phase Deposition

Organic vapor phase deposition (OVPD) allows better control of the structure and morphology of the film than vacuum thermal evaporation. The process involves evaporation of the organic material over a substrate in the presence of an inert carrier gas. The resulting film morphology can be tuned by changing the gas flow rate and the source temperature. Uniform films can be grown by reducing the carrier gas pressure, which will increase the velocity and mean free path of the gas, and as a result boundary layer thickness decreases. Cells produced by OVPD do not have issues related with contaminations from the flakes coming out of the walls of the chamber, as the walls are warm and do not allow molecules to stick to and produce a film upon them.

Another advantage over VTE is the uniformity in evaporation rate. This occurs because the carrier gas becomes saturated with the vapors of the organic material coming out of the source and then moves towards the cooled substrate. Depending on the growth parameters (temperature of the source, base pressure and flux of the carrier gas) the deposited film can be crystalline or amorphous in nature. Devices fabricated using OVPD show a higher short-circuit current density than that of devices made using VTE. An extra layer of donor-acceptor hetero-junction at the top of the cell may block excitons, whilst allowing conduction of electron; resulting in improved cell efficiency.

Organic Solar Ink

Plextronics Plexcore PV 2000 organic solar ink is able to deliver higher performance in *fluorescent* lighting conditions in comparison to amorphous silicon solar cells, and said to have a 30% to 40% increase in indoor power density in comparison to the standard organic solar technology. Plextronics has also developed a manufacturing method that allows for low-temperature processing of OPV. While previous industry standard techniques required a glass substrate to be annealed at 110 °C or higher temperatures, this method enables annealing below 65 °C. It is expected to reduce manufacturing costs as it may enable the use of less expensive substrates, especially once the process is transferred to flexible substrates such as plastic.

Conductive Ink

'Conductive ink' is an ink that results in a printed object which conducts electricity. The transformation from liquid ink to solid printing may involve drying, curing or melting processes.

These inks may be classed as fired high solids systems or PTF polymer thick film systems that allow circuits to be drawn or printed on a variety of substrate materials such as polyester to paper. These types of inks usually contain conductive materials such as powdered or flaked silver and carbon like materials, although polymeric conduction is also known.

Conductive inks can be a more economical way to lay down a modern conductive traces when compared to traditional industrial standards such as etching copper from copper plated substrates to form the same conductive traces on relevant substrates, as printing is a purely additive process producing little to no waste streams which then have to be recovered or treated.

Silver inks have multiple uses today including printing RFID tags as used in modern transit tickets, they can be used to improvise or repair circuits on printed circuit boards. Computer keyboards contain membranes with printed circuits that sense when a key is pressed. Windshield defrosters consisting of resistive traces applied to the glass are also printed. Many newer cars have conductive traces printed on a rear window, serving as the radio antenna.

Printed paper and plastic sheets have problematic characteristics, primarily high resistance and lack of rigidity. The resistances are too high for the majority of circuit board work, and the non-rigid nature of the materials permits undesirable forces to be exerted on component connections, causing reliability problems. Consequently such materials are only used in a restricted range of applications, usually where the flexibility is important and no parts are mounted on the sheet.

Mix 75.5 parts gallium with 24.5 parts indium in a beaker of deionized water, heat to 50 °C, stir, and voilα: an alloy that's liquid at room temperature, costs about US $1 per milliliter, and is two orders of magnitude more conductive than the nanoparticle inks; its resistivity is just 17 times that of copper.

Low Cost Solar Cell

A new type of solar cell, made from a material that is dramatically cheaper to obtain and use than silicon, could generate as much power as today's commodity solar cells. Although the potential of the material is just starting to be understood, it has caught the attention of the world's leading solar researchers, and several companies are already working to commercialize it.

Researchers developing the technology say that it could lead to solar panels that cost just 10 to 20 cents per watt. Solar panels now typically cost about 75 cents a watt. In the past, solar researchers have been divided into two camps in their pursuit of cheaper solar power. Some have sought solar cells that can be made very cheaply but that have the downside of being relatively inefficient. Lately, more researchers have focused on developing very high efficiency cells, even if they require more expensive manufacturing techniques.

The new material may make it possible to get the best of both worlds—solar cells that are highly efficient but also cheap to make. Perovskites have been known for over a century, but no one thought to try them in solar cells until relatively recently. The particular material the researchers are using is very good at absorbing light. While conventional silicon solar panels use materials that are about 180 micrometres thick, the new solar cells use less than one micrometre of material to capture the same amount of sunlight. The pigment is a semiconductor that is also good at transporting the electric charge created when light hits it.

The efficiencies of mass-produced solar cells may be lower. But it makes sense to compare the lab efficiencies of the perovskite cells with the lab records for other materials. Grätzel says that perovskite in solar cells will likely prove to be a "forgiving" material that retains high efficiencies in mass production, since the manufacturing processes are simple.

Perovskite solar cells can be made by spreading the pigment on a sheet of glass or metal foil, along with a few other layers of material that facilitate the movement of electrons through the cell. This isn't quite the spray-on solar cells that some people have envisioned—a sci-fi ideal of instantly converting any surface into one that can generate electricity—but the process is so easy that it's getting close. When perovskites were first tried in solar cells in 2009, efficiencies were low—they only converted about 3.5 percent of the energy in sunlight into electricity. The cells also didn't last very long, since a liquid electrolyte dissolved the perovskite. Researchers barely had enough time to test them before they stopped working. But last year a couple of technical innovations—ways to replace a liquid electrolyte with solid materials—solved those problems and started researchers on a race to produce ever-more-efficient solar cells.

Like any other new entrant into the highly competitive solar-panel market, perovskites will have difficulty taking on silicon solar cells. The costs of silicon solar cells are falling, and some analysts think they could eventually fall as low as 25 cents per watt, which would eliminate most of the cost advantage of perovskites and lessen the incentive for investing in the new technology. The manufacturing process for perovskite solar cells—which can be as simple as spreading a liquid over a surface or can involve vapor deposition, another large-scale manufacturing process—is expected to be easy. But historically, it has taken over a decade to scale up novel solar-cell technologies, and a decade from now silicon solar cells could be too far ahead to catch.

Green says one opportunity may be to use perovskites to augment rather than replace silicon solar cells. It might be possible to paint perovskites onto conventional silicon solar cells to improve their efficiency, and so lower the overall cost per watt for solar cells. This might be an easier way to break into the solar market than trying to introduce an entirely new kind of solar cell.

A challenge may be the fact that the material includes a small amount of lead, which is toxic. Tests will be needed to show how toxic it is as part of the perovskite material. Steps can also be taken to ensure the solar cells are collected and recycled to prevent the materials from getting into the environment—the approach pursued now with the lead-acid starter batteries used in cars. It may also be possible to substitute tin or some other element for lead in the cells.

Multijunction Photovoltaic Cell

Multi-junction solar cells are solar cells with multiple p-n junctions made of different semiconductor materials. Each material's p-n junction will produce electric current in response to a different wavelength of light. A multi-junction cell solar cell will produce electric current at multiple wavelengths of light, increasing the conversion efficiency of the no-cost solar light power to usable electric power.

Traditional single-junction cells have a maximum theoretical efficiency of 34%. Theoretically, with an infinite number of junctions, multi-junction cell efficiency would be 87% under highly concentrated sunlight.

Currently, the best lab examples of traditional silicon solar cells have efficiencies around 25%, while lab examples of multi-junction cells have demonstrated performance over 43%. Commercial examples of tandem, two layer, cells are widely available at 30% under one-sun illumination, and improve to around 40% under concentrated sunlight. However, this efficiency is gained at the cost of increased complexity and manufacturing price. To date, their higher price and higher price-to-performance ratio have limited their use to special roles, notably in aerospace where their high power-to-weight ratio is desirable. In terrestrial applications these solar cells have been suggested for use in concentrated photovoltaics (CPV), with numerous small test sites around the world.

Tandem fabrication techniques are used to improve the performance of existing many designs. In particular, the technique can be applied to lower cost thin-film solar cells using amorphous silicon, as opposed to mono-crystalline silicon, to produce a cell with about 10% efficiency that is lightweight and flexible. This approach has been used by several commercial vendors, but these products are currently limited to certain niche roles, like roofing materials.

Description

Basics of Solar Cells: Traditional photovoltaic cells are commonly composed by doped silicon and then depositing metallic contacts on the top and bottom. The doping is normally applied to a thin layer on the top of the cell, producing a pn-junction with a particular bandgap energy, E_g.

Photons that hit the top of the solar cell are either reflected or transmitted into the cell. Transmitted photons have the potential to give their energy *h*n to an electron if *h*n ³ E_g, generating an electron-hole pair. In the depletion region, the drift electric field E_{drift} accelerates both electrons and holes towards their respective n-doped and p-doped regions (up and down, respectively). The resulting current I_g is called the generation photocurrent. In the quasi-neutral region, the scattering electric field E_{scatt} accelerates holes (electrons) towards the p-doped (n-doped) region, which gives a scattering photocurrent I_{pscatt} (I_{nscatt}). Consequently, due to the accumulation of charges, a potential V and a photocurrent I_{ph} appear. The expression for this photocurrent is obtained by adding generation and scattering photocurrents: $I_{ph} = I_g + I_{nscatt} + I_{pscatt}$.

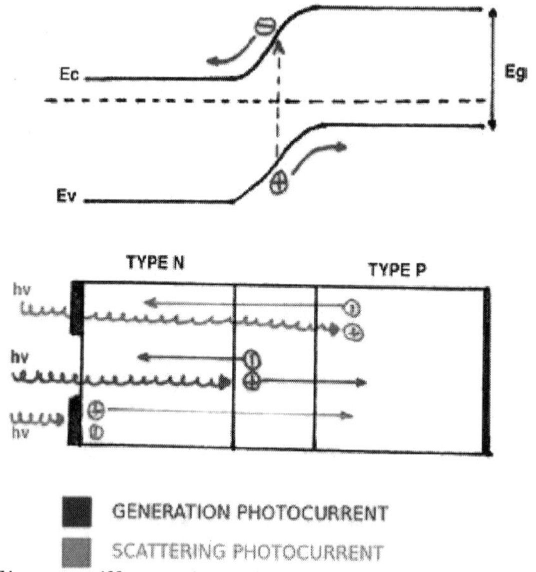

Figure: *A. Band diagram illustration of the photovoltaic effect. Photons give their energy to electrons in the depletion or quasi-neutral regions. These move from conduction band to valence band. Depending on the location, electrons and holes are accelerated by* E_{drift}, *which gives generation photocurrent, or by* E_{scatt}, *which gives scattering photocurrent.*

The *J-V* characteristics (J is current density, i.e. current per unit area) of a solar cell under illumination are obtained by shifting the *J-V* characteristics of a diode in the dark downward by I_{ph}. Since solar cells are designed to supply power and not absorb it, the power $P = V \cdot I_{ph}$ must be negative. Hence, the operating point (V_m, J_m) is located in the region where $V>0$ and $I_{ph}<0$, and chosen to maximize the absolute value of the power $|P|$.

Loss Mechanisms

The theoretical performance of a solar cell was first studied in depth in the 1960s, and is today known as the Shockley–Queisser limit. The limit describes several loss mechanisms that are inherent to any solar cell design. The first are the losses due to blackbody radiation, a loss mechanism that affects any material object above absolute zero. In the case of solar cells at standard temperature and pressure, this loss accounts for about 7% of the power. The second is an effect known as "recombination", where the electrons created by the photoelectric effect meet the electron holes left behind by previous excitations. In silicon, this accounts for another 10% of the power.

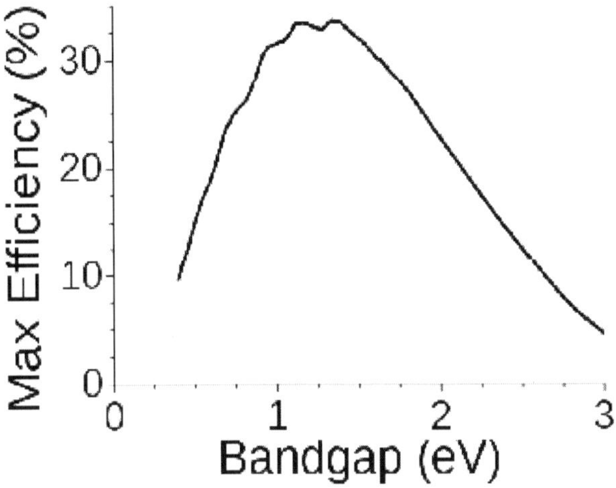

Figure: *The Shockley-Queisser limit for the efficiency of a single-junction solar cell. It is essentially impossible for a single-junction solar cell, under unconcentrated sunlight, to have more than ~34% efficiency. A multijunction cell, however, can exceed that limit.*

However, the dominant loss mechanism is the inability for a solar cell to extract all of the power in the photon, and the associated problem that it cannot extract any power at all from certain photons. This is due to the fact that the electrons must have enough energy to overcome the bandgap of the material.

If the photon has less energy than the bandgap, it is not collected at all. This is a major consideration for conventional solar cells, which are not sensitive to most of the infrared spectrum, although that represents almost half of the power coming from the sun. Conversely, photons with more energy than the bandgap, say blue light, initially eject an electron to a state high above the bandgap, but this extra energy is lost through collisions in a process known as "relaxation". This lost energy turns into heat in the cell, which has the side-effect of further increasing blackbody losses.

Combining all of these factors, the maximum efficiency for a single-bandgap material, like conventional silicon cells, is about 34%. That is, 66% of the energy in the sunlight hitting the cell will be lost. Practical concerns further reduce this, notably reflection off the front surface or the metal terminals, with modern high-quality cells at about 22%.

Lower, also called narrower, bandgap materials will convert longer wavelength, lower energy photons. Higher, or wider bandgap materials will convert shorter wavelength, higher energy light An analysis of the AM1.5 spectrum, shows the best balance is reached at about 1.1 eV, in the near infrared, which happens to be very close to the natural bandgap in silicon and a number of other useful semiconductors.

Multi-Junction Cells

Cells made from multiple materials have multiple bandgaps. So, it will respond to multiple light wavelengths and some of the energy that would otherwise be lost to relaxation as described above, can be captured and converted.

For instance, if one had a cell with two bandgaps in it, one tuned to red light and the other to green, then the extra energy in green, cyan and blue light would be lost only to the bandgap of the green-sensitive material, while the energy of the red, yellow and orange would be lost only to the bandgap of the red-sensitive material. Following analysis similar to those performed for single-bandgap devices, it can be demonstrated that the perfect bandgaps for a two-gap device are at 1.1 eV and 1.8 eV.

Conveniently, light of a particular wavelength does not interact strongly with materials that are not a multiple of that wavelength. This means that you can make a multijunction cell by layering the different materials on top of each other, shortest wavelengths on the "top" and increasing through the body of the cell. As the photons have to pass through the cell to reach the proper layer to be absorbed, transparent conductors need to be used to collect the electrons being generated at each layer.

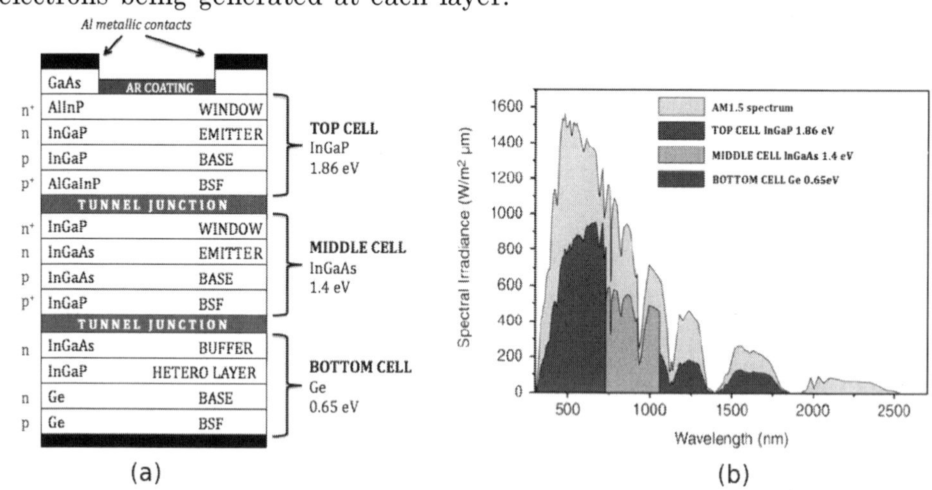

Figure: *C. (a) The structure of an MJ solar cell. There are six important types of layers: pn junctions, back surface field (BSF) layers, window layers, tunnel junctions, anti-reflective coating and metallic contacts. (b) Graph of spectral irradiance E vs. wavelength λ over the AM1.5 solar spectrum, together with the maximum electricity conversion efficiency for every junction as a function of the wavelength.*

Producing a tandem cell is not an easy task, largely due to the thinness of the materials and the difficulties extracting the current between the layers. The easy solution is to use two mechanically separate thin film solar cells and then wire them together separately outside the cell. This technique is widely used by amorphous silicon solar cells, Uni-Solar's products use three such layers to reach efficiencies around 9%. Lab examples using more exotic thin-film materials have demonstrated efficiencies over 30%.

The more difficult solution is the "monolithically integrated" cell, where the cell consists of a number of layers that are mechanically and electrically connected. These cells are much more difficult to produce because the electrical characteristics of each layer has to be carefully matched. In particular, the photocurrent generated in each layer needs to be matched, otherwise electrons will be absorbed between layers. This limits their construction to certain materials, best met by the III-V semiconductors.

Material Choice

The choice of materials for each sub-cell is determined by the requirements for lattice-matching, current-matching, and high performance opto-electronic properties. For optimal growth and resulting crystal quality, the crystal lattice constant *a* of each material must be closely matched, resulting in lattice-matched devices.

This constraint has been relaxed somewhat in recently developed metamorphic solar cells which contain a small degree of lattice mismatch. However, a greater degree of mismatch or other growth imperfections can lead to crystal defects causing a degradation in electronic properties.

Since each sub-cell is connected electrically in series, the same current flows through each junction. The materials are ordered with decreasing bandgaps, E_g, allowing sub-bandgap light ($hc/\lambda < e \cdot E_g$) to transmit to the lower sub-cells. Therefore, suitable bandgaps must be chosen such that the design spectrum will balance the current generation in each of the sub-cells, achieving current matching. It is plotted together with the maximum conversion efficiency for every junction as a function of the wavelength, which is directly related to the number of photons available for conversion into photocurrent.

Finally, the layers must be electrically optimal for high performance. This necessitates usage of materials with strong absorption coefficients $\alpha(\lambda)$, high minority carrier lifetimes $\tau_{minority}$, and high mobilities μ.

The favourable values in the table below justify the choice of materials typically used for multi-junction solar cells: InGaP for the top sub-cell (E_g = 1.8 - 1.9 eV), InGaAs for the middle sub-cell (E_g = 1.4 eV), and Germanium for the bottom sub-cell (E_g = 0.67 eV). The use of Ge is mainly due to its lattice constant, robustness, low cost, abundance, and ease of production.

Because the different layers are closely lattice-matched, the fabrication of the device typically employs metal-organic chemical vapor deposition (MOCVD). This technique is preferable to the molecular beam epitaxy (MBE) because it ensures high crystal quality and large scale production.

Material	E_g, *eV*	*a, nm*	*absorption (λ = 0.8 μm), 1/μm*	μ_n, *cm²/(V·s)*	τ_p, *μs*	*Hardness (Mohs)*	*α, μm/K*	*S, m/s*
c-Si	1.12	0.5431	0.102	1400	1	7	2.6	0.1–60
InGaP	1.86	0.5451	2	500	–	5	5.3	50
GaAs	1.4	0.5653	0.9	8500	3	4–5	6	50
Ge	0.65	0.5657	3	3900	1000	6	7	1000
InGaAs	1.2	0.5868	30	1200	–	–	5.66	100–1000

Structural Elements

Metallic Contacts

The metallic contacts are low-resistivity electrodes that make contact with the semiconductor layers. They are often aluminum. This provides an electrical connection to a load or other parts of a solar cell array. They are usually on two sides of the cell. And are can be important on the back face of so that shadowing on the lighting surface is reduced.

Anti-Reflective Coating

Anti-reflective (AR) coating is generally composed of several layers in the case of MJ solar cells. The top AR layer has usually a NaOH surface texturation with several pyramids in order to increase the transmission coefficient *T*, the trapping of the light in the material (because photons cannot easily get out the MJ structure due to pyramids) and therefore, the path length of photons in the material. On the one hand, the thickness of each AR layer is chosen to get destructive interferences. Therefore, the reflection coefficient *R* decreases to 1%. In the case of two AR layers L_1 (the top layer, usually SiO_2) and L_2 (usually TiO_2), there must be $n_{L2} = n_{AlInP}^{1/2} \cdot n_{L1}$ to have the same amplitudes for reflected fields and $n_{L1}d_{L1} = 4\lambda_{min}, n_{L2}d_{L2} = \lambda_{min}/4$ to have opposite phase for reflected fields. On the other hand, the thickness of each AR layer is also chosen to minimize the reflectance at wavelengths for which the photocurrent is the lowest. Consequently, this maximizes J_{SC} by matching currents of the three subcells. As example, because the current generated by the bottom cell is greater than the currents generated by the other cells, the thickness of AR layers is adjusted so that the infrared (IR) transmission (which corresponds to the bottom cell) is degraded while the ultraviolet transmission (which corresponds to the top cell) is upgraded. Particularly, an AR coating is very important at low wavelengths because, without it, *T* would be strongly reduced to 70%.

Tunnel Junctions

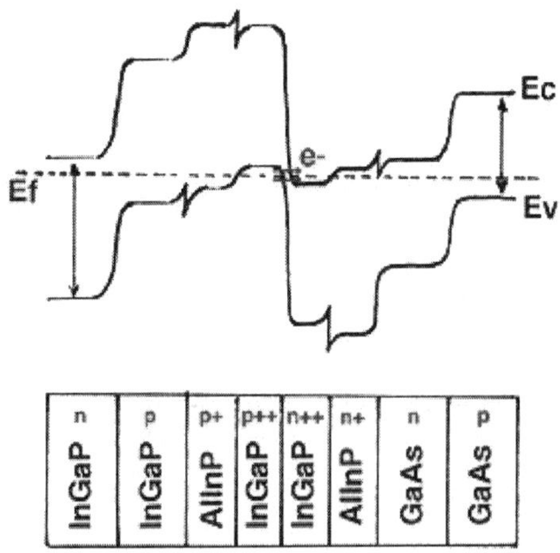

Figure D: *Layers and band diagram of the tunnel junction. Because the length of the depletion region is narrow and the band gap is high, electrons can tunnel.*

The main goal of tunnel junctions is to provide a low electrical resistance and optically low-loss connection between two subcells. Without it, the p-doped region of the top cell would be directly connected with the n-doped region of the middle cell. Hence, a pn junction with opposite direction to the others would appear between the top cell and the middle cell. Consequently, the photovoltage would be lower than if there would be no parasitic diode. In order to decrease this effect, a tunnel junction is used. It is simply a wide band gap, highly doped diode. The high doping reduces the length of the depletion region because

$$l_{depl} = \sqrt{\frac{2\epsilon(\phi_0 - V)}{q}\frac{N_A + N_D}{N_A N_D}}$$

Hence, electrons can easily tunnel through the depletion region. The J-V characteristic of the tunnel junction is very important because it explains why tunnel junctions can be used to have a low electrical resistance connection between two pn junctions. Figure shows three different regions: the tunneling region, the negative differential resistance region and the thermal diffusion region. The region where electrons can tunnel through the barrier is called the tunneling region. There, the voltage must be low enough so that energy of some electrons who are tunneling is equal to energy states available on the other side of the barrier. Consequently, current density through the tunnel junction is high (with maximum value of J_P, the peak current density) and the slope near the origin is therefore steep. Then, the resistance is extremely low and consequently, the voltage too. This is why tunnel junctions are ideal for connecting two pn junctions without having a voltage drop. When voltage is higher, electrons cannot cross the barrier because energy states are no longer available for electrons. Therefore, the current density decreases and the differential resistance is negative. The last region, called thermal diffusion region, corresponds to the J-V characteristic of the usual diode:

$$J = J_S\left(\exp\left(\frac{qV}{kT}\right) - 1\right)$$

In order to avoid the reduction of the MJ solar cell performances, tunnel junctions must be transparent to wavelengths absorbed by the next photovoltaic cell, the middle cell, i.e. $E_{gTunnel} > E_{gMiddleCell}$.

Window Layer and Back-Surface Field

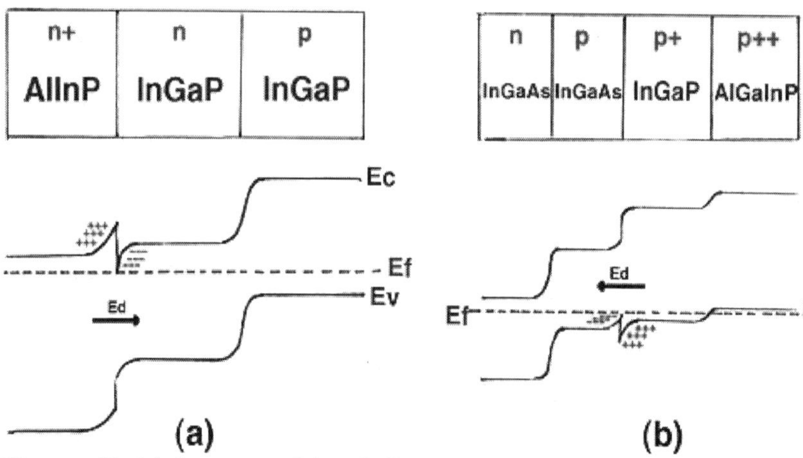

Figure: E: (a) Layers and band diagram of a window layer. The surface recombination is reduced. (b) Layers and band diagram of a BSF layer. The scattering of carriers is reduced.

A window layer is used in order to reduce the surface recombination velocity *S*. Similarly, a back-surface field (BSF) layer reduces the scattering of carriers towards the tunnel junction.

The structure of these two layers is the same: it is a heterojunction which catches electrons (holes). Indeed, despite the electric field E_d, these cannot jump above the barrier formed by the heterojunction because they don't have enough energy. Hence, electrons (holes) cannot recombine with holes (electrons) and cannot diffuse through the barrier. By the way, window and BSF layers must be transparent to wavelengths absorbed by the next pn junction i.e. $E_{gWindow} > E_{gEmitter}$ and $E_{gBSF} > E_{gEmitter}$. Furthermore, the lattice constant must be close to the one of InGaP and the layer must be highly doped ($n \geq 10^{18}$ cm^{-3}).

Materials

The majority of multi-junction cells that have been produced to date use three layers. These cells require the use of semiconductors that can be tuned to specific frequencies, which has led to most of them being made of gallium arsenide (GaAs) compounds, often germanium for the bottom-, GaAs for the middle-, and $GaInP_2$ for the top-cell.

Gallium Arsenide Substrate

Dual junction cells can be made on Gallium arsenide wafers. Alloys of Indium gallium phosphide in the range $In_{.5}Ga_{.5}P$ through $In_{.53}Ga_{.47}P$ serve as the high band gap alloy. This alloy range provides for the ability to have band gaps in the range of 1.92eV to 1.87eV. The lower GaAs junction has a band gap of 1.42eV.

Germanium Substrate

Triple junction cells consisting of Indium gallium phosphide, Gallium arsenide or Indium gallium arsenide and Germanium can be fabricated on germanium wafers. Early cells used straight gallium arsenide in the middle junction. Later cells have utilised $In_{0.015}Ga_{0.985}As$, due to the better lattice match to Ge, resulting in a lower defect density.

Due to the huge band gap difference between GaAs (1.42eV), and Ge (0.66eV), the current match is very poor, with the Ge junction operated significantly current limited.

Current efficiencies for commercial InGaP/GaAs/Ge cells approach 40% under concentrated sunlight . Lab cells (partly using additional junctions between the GaAs and Ge junction) have demonstrated efficiencies above 40%.

Indium Phosphide Substrate

Indium phosphide may be used as a substrate to fabricate cells with band gaps between 1.35eV and 0.74eV. Indium Phosphide has a band gap of 1.35eV. Indium gallium arsenide($In_{0.53}Ga_{0.47}As$) is lattice matched to Indium Phosphide with a band gap of 0.74eV. A quaternary alloy of Indium gallium arsenide phosphide can be lattice matched for any band gap in between the two.

Indium phosphide-based cells have the potential to work in tandem with gallium arsenide cells. The two cells can be optically connected in series (with the InP cell below the GaAs cell), or in parallel through the use of spectra splitting using a Dichroic filter.

Performance Improvements

Structure

All MJ photovoltaic cells use III-V semiconductor materials. GaAsSb-based heterojunction tunnel diodes, instead of conventional InGaP highly doped tunnel diodes described above, have a lower tunneling distance. Indeed, in the heterostructure formed by GaAsSb and InGaAs, the valence band of GaAsSb is higher than the valence band of the adjoining p-doped layer.

Consequently, the tunneling distance d_{tunnel} is reduced and so the tunneling current, which exponentially depends of d_{tunnel}, is increased. Hence, the voltage is lower than that of the InGaP tunnel junction. GaAsSb heterojunction tunnel diodes offer other advantages.

The same current can be achieved by using a lower doping. Secondly, because the lattice constant is larger for GaAsSb than Ge, one can use a wider range of materials for the bottom cell because more materials are lattice-matched to GaAsSb than to Ge.

Chemical components can be added to some layers. Adding about one percent of Indium in each layer better matches lattice constants of the different layers. Without it, there is about 0.08 percent of mismatching between layers, which inhibits performance. Adding aluminium to the top cell increases its band gap to 1.96 eV, covering a larger part of the solar spectrum and obtain a higher open-circuit voltage V_{OC}.

The theoretical efficiency of MJ solar cells is 86.8% for an infinite number of pn junctions, implying that more junctions increase efficiency. The maximum theoretical efficiency is 37, 50, 56, 72% for 1, 2, 3, 36 pn junctions, respectively, with the number of junctions increasing exponentially to achieve equal efficiency increments.

The exponential relationship implies that as the cell approaches the limit of efficiency, the increase cost and complexity grow rapidly. Decreasing the thickness of the top cell increases the transmission coefficient T.

Finally, an InGaP hetero-layer between the p-Ge layer and the InGaAs layer can be added in order to create automatically the n-Ge layer by scattering during MOCVD growth and increase significantly the quantum efficiency $QE(\lambda)$ of the bottom cell. InGaP is advantageous because of its high scattering coefficient and low solubility in Ge.

Spectral Variations

Solar spectrum at the Earth surface changes constantly depending on the weather and sun position. This results in the variation of $\varphi(\lambda)$, $QE(\lambda)$, $\alpha(\lambda)$ and thus the short-circuit currents J_{SCi}. As a result, the current densities J_i are not necessarily matched and the total current becomes lower. These variations can be quantified using the average photon energy (APE) which is the ratio between the spectral irradiance $G(\lambda)$ (the power density of the light source in a specific wavelength λ) and the total photon flux density. It can be shown that a high (low) value for APE means low (high) wavelengths spectral conditions and higher (lower) efficiencies.

Thus APE is a good indicator for quantifying the effects of the solar spectrum variations on performances and has the added advantage of being independent of the device structure and the absorption profile of the device.

Use of Light Concentrators

Light concentrators increase efficiencies and reduce the cost/efficiency ratio. The three types of light concentrators in use are refractive lenses like Fresnel lenses, reflective dishes (parabolic or cassegraine), and light guide optics. Thanks to these devices, light arriving on a large surface can be concentrated on a smaller cell.

The intensity concentration ratio (or "suns") is the average intensity of the focused light divided by 0.1 W/cm^2. If its value is X then the MJ current becomes X higher under concentrated illumination.

Using concentrations on the order of 500 to 1000, meaning that a 1 cm^2 cell can use the light collected from 0.1 m^2 (as 1 m^2 equal 10000 cm^2), produces the highest efficiencies seen to date. Three-layer cells are fundamentally limited to 63%, but existing commercial prototypes have already demonstrated over 40%. These cells capture about 2/3 of their theoretical maximum performance, so assuming the same is true for a non-concentrated version of the same design, one might expect a three-layer cell of 30% efficiency. This is not enough of an advantage over traditional silicon designs to make up for their extra production costs. For this reason, almost all multi-junction cell research for terrestrial use is dedicated to concentrator systems, normally using mirrors or fresnel lenses.

Using a concentrator also has the added benefit that the number of cells needed to cover a given amount of ground area is greatly reduced. A conventional system covering 1 m^2 would require 625 16 cm^2 cells, but for a concentrator system only a single cell is needed, along with a concentrator.

The argument for concentrated Multi-junction cells has been that the high cost of the cells themselves would be more than offset by the reduction in total number of cells. However, the downside of the concentrator approach is that efficiency drops off very quickly under lower lighting conditions. In order to maximize its advantage over traditional cells and thus be cost competitive, the concentrator system has to track the sun as it moves to keep the light focused on the cell and maintain maximum efficiency as long as possible. This requires a solar tracker system, which increases yield, but also cost.

Fabrication

Multi-junction cells are expensive to produce, using techniques similar to semiconductor device fabrication, usually metalorganic vapour phase epitaxy but on "chip" sizes on the order of centimetres. In cases where outright performance is the only consideration, these cells have become common, they are widely used in satellite applications for instance, where the power-to-weight ratio overwhelms practically every other cost.

Comparison with Other Technologies

There are four main categories of photovoltaic cells: c-Si solar cells, thin film solar cells, MJ solar cells and new technologies (including organic solar cells).

Technology	h (%)	V_{OC} (V)	I_{SC} (A)	W/m²	t (µm)
u c-Si	24.7	0.5	0.8	63	100
p c-Si	20.3	0.615	8.35	211	200
a-Si	11.1	6.3	0.0089	33	1
CdTe	16.5	0.86	0.029	–	5
CIGS	19.5	–	–	–	1
MJ	40.7	2.6	1.81	476	140

MJ solar cells and other photovoltaic devices have significant differences. Physically, the main property of a MJ solar cell is having more than one pn junction in order to catch a larger photon energy spectrum while the main property of the thin film solar cell is to use thin films instead of thick layers in order to decrease the cost efficiency ratio. As of 2010, MJ solar panels are more expensive than others. These differences imply different applications: MJ solar cells are preferred in space and c-Si solar cells for terrestrian applications.

6

Solar Cells

A solar cell (also called a photovoltaic cell) is an electrical device that converts the energy of light directly into electricity by the photovoltaic effect. It is a form of photoelectric cell (in that its electrical characteristics—e.g. current, voltage, or resistance—vary when light is incident upon it) which, when exposed to light, can generate and support an electric current without being attached to any external voltage source, but do require an external load for power consumption.

The term "photovoltaic" comes from the Greek φφὸ (*phτs*) meaning "light", and from "volt", the unit of electro-motive force, the volt, which in turn comes from the last name of the Italian physicist Alessandro Volta, inventor of the battery (electrochemical cell). The term "photo-voltaic" has been in use in English since 1849.

Photovoltaics is the field of technology and research related to the practical application of photovoltaic cells in producing electricity from light, though it is often used specifically to refer to the generation of electricity from sunlight. Cells can be described as *photovoltaic* even when the light source is not necessarily sunlight (lamplight, artificial light, etc.). In such cases the cell is sometimes used as a photodetector (for example infrared detectors), detecting light or other electromagnetic radiation near the visible range, or measuring light intensity.

The operation of a photovoltaic (PV) cell requires 3 basic attributes:

1. The absorption of light, generating either electron-hole pairs or excitons.
2. The separation of charge carriers of opposite types.
3. The separate extraction of those carriers to an external circuit.

In contrast, a solar thermal collector supplies heat by absorbing sunlight, for the purpose of either direct heating or indirect electrical power generation. "Photoelectrolytic cell" (photoelectrochemical cell), on the other hand, refers either to a type of photovoltaic cell (like that developed by A.E. Becquerel and modern dye-sensitized solar cells), or to a device that splits water directly into hydrogen and oxygen using only solar illumination.

Building Block of a Solar Panel

Assemblies of photovoltaic cells are used to make solar modules which generate electrical power from sunlight. Multiple cells in an integrated group, all oriented in one plane, constitute a *solar photovoltaic panel* or "solar photovoltaic module," as distinguished from a "solar thermal module" or "solar hot water panel." The electrical energy generated from solar modules, colloquially referred to as *solar power*, is an example of *solar energy*. A group of connected solar modules is called an "array."

History of Solar Cells

The photovoltaic effect was first experimentally demonstrated by French physicist A. E. Becquerel. In 1839, at age 19, experimenting in his father's laboratory, he built the world's first photovoltaic cell. Willoughby Smith first described the "Effect of Light on Selenium during the passage of an Electric Current" in an article that was published in the 20 February 1873 issue of Nature. However, it was not until 1883 that the first solid state photovoltaic cell was built, by Charles Fritts, who coated the semiconductor selenium with an extremely thin layer of gold to form the junctions. The device was only around 1% efficient. In 1888 Russian physicist Aleksandr Stoletov built the first cell based on the outer photoelectric effect discovered by Heinrich Hertz earlier in 1887.

Albert Einstein explained the underlying mechanism of light instigated carrier excitation—the photoelectric effect—in 1905, for which he received the Nobel prize in Physics in 1921. Russell Ohl patented the modern junction semiconductor solar cell in 1946, which was discovered while working on the series of advances that would lead to the transistor.

The first practical photovoltaic cell was developed in 1954 at Bell Laboratories by Daryl Chapin, Calvin Souther Fuller and Gerald Pearson. They used a diffused silicon p–n junction that reached 6% efficiency, compared to the selenium cells that found it difficult to reach 0.5%. Les Hoffman CEO of Hoffman Electronics Corporation had his Semiconductor Division pioneer the fabrication and mass production of solar cells. From 1954 to 1960 Hoffman improved the efficiency of Solar Cells from 2% to 14%. At first, cells were developed for toys and other minor uses, as the cost of the electricity they produced was very high; in relative terms, a cell that produced 1 watt of electrical power in bright sunlight cost about $250, comparing to $2 to $3 per watt for a coal plant.

Solar cells were brought from obscurity by the suggestion to add them, probably due to the successes made by Hoffman Electronics, to the Vanguard I satellite, launched in 1958. In the original plans, the satellite would be powered only by battery, and last a short time while this ran down. By adding cells to the outside of the body, the mission time could be extended with no major changes to the spacecraft or its power systems. In 1959 the United States launched Explorer 6. It featured large solar arrays resembling wings, which became a common feature in future satellites. These arrays consisted of 9600 Hoffman solar cells. There was some skepticism at first, but in practice the cells proved to be a huge success, and solar cells were quickly designed into many new satellites, notably Bell's own Telstar.

Improvements were slow over the next two decades, and the only widespread use was in space applications where their power-to-weight ratio was higher than any competing technology. However, this success was also the reason for slow progress; space users were willing to pay anything for the best possible cells, there was no reason to invest in lower-cost solutions if this would reduce efficiency. Instead, the price of cells was determined largely by the semiconductor industry; their move to integrated circuits in the 1960s led to the availability of larger boules at lower relative prices. As their price fell, the price of the resulting cells did as well. However these effects were limited, and by 1971 cell costs were estimated to be $100 per watt.

Berman's Price Reductions

In the late 1960s, Elliot Berman was investigating organic solar cells. He was introduced to a team at Exxon who were looking for projects 30 years in the future. The group had concluded that electrical power would be much more expensive by 2000, and felt that this increase in price would make new alternative energy sources more attractive, and solar was the most interesting among these. In 1969, Berman joined the Linden, New Jersey Exxon lab, Solar Power Corporation (SPC).

His first major effort was to canvas the potential market to see what possible uses for a new product were, and they quickly found that if the price per watt were reduced from then-current $100/watt to about $20/watt there would be significant demand. Knowing that his ribbon concept would take years to develop, the team started looking for ways to hit the $20 price point using existing materials. The first improvement was the realisation that the existing cells were based on standard semiconductor manufacturing process, even though that was not ideal. This started with the boule, cutting it into disks called wafers, polishing the wafers, and then, for cell use, coating them with an anti-reflective layer. Berman noted that the rough-sawn wafers already had a perfectly suitable anti-reflective front surface, and by printing the electrodes directly on this surface, two major steps in the cell processing were eliminated. The team also explored ways to improve the mounting of the cells into arrays, eliminating the expensive materials and hand wiring used in space applications. Their solution was to use a printed circuit board on the back, acrylic plastic on the front, and silicone glue between the two, potting the cells. The largest improvement in price point was Berman's realisation that existing silicon was effectively "too good" for solar cell use; the minor imperfections that would ruin a boule (or individual wafer) for electronics would have little effect in the solar application. Solar cells could be made using cast-off material from the electronics market.

Putting all of these changes into practice, the company started buying up "reject" silicon from existing manufacturers at very low cost. By using the largest wafers available, thereby reducing the amount of wiring for a given panel area, and packaging them into panels using their new methods, by 1973 SPC was producing panels at $10 per watt and selling them at $20 per watt, a fivefold decrease in prices in two years.

Navigation Market

SPC approached companies making navigational buoys as a natural market for their products, but found a curious situation. The primary company in the business was Automatic Power, a battery manufacturer. Realising that solar cells might eat into their battery profits, Automatic had purchased a solar navigation aid prototype from Hoffman Electronics and shelved it. Seeing there was no interest at Automatic Power, SPC turned to Tideland Signal, another battery company formed by ex-Automatic managers. Tideland introduced a solar-powered buoy and was soon ruining Automatic's business.

The timing could not be better; the rapid increase in the number of offshore oil platforms and loading facilities produced an enormous market among the oil companies. As Tideland's fortunes improved, Automatic Power started looking for their own supply of solar panels. They found Bill Yerks of Solar Power International (SPI) in California, who was looking for a market. SPI was soon bought out by one of its largest customers, the ARCO oil giant, forming ARCO Solar. ARCO Solar's factory in Camarillo, California was the first dedicated to building solar panels, and was in continual operation from its purchase by ARCO in 1977 until 2011 when it was closed by SolarWorld.

This market combined with the 1973 oil crisis. Oil companies were now cash-flush due to their huge profits during the crisis, but were also acutely aware that their future success would depend on some other form of power. Over the next few years, major oil companies started a number of solar firms, and were for decades the largest producers of solar panels. Exxon, ARCO, Shell, Amoco (later purchased by BP) and Mobil all had major solar divisions during the 1970s and 1980s. Technology companies also had some investment, including General Electric, Motorola, IBM, Tyco and RCA.

Further Improvements

In recent times, further improvements have brought production costs down under $1 a watt, with wholesale costs well under $2. "Balance of system" costs are now more than the panels themselves. Large commercial arrays can be built, as of 2010, at below $3.40 a watt, fully commissioned.

As the semiconductor industry moved to ever-larger boules, older equipment became available at fire-sale prices. Cells have grown in size as older equipment became available on the surplus market; ARCO Solar's original panels used cells with 2 to 4 inch (50 to 100 mm) diameter. Panels in the 1990s and early 2000s generally used 5 inch (125 mm) wafers, and since 2008 almost all new panels use 6 inch (150 mm) cells. The widespread introduction of flat screen televisions in the late 1990s and early 2000s led to the wide availability of large sheets of high-quality glass, used on the front of the panels.

During the 1990s, polysilicon cells became increasingly popular. These cells offer less efficiency than their monosilicon counterparts, but they are grown in large vats that greatly reduce the cost of production. By the mid-2000s, poly was dominant in the low-cost panel market, but more recently a variety of factors has pushed the higher performance mono back into widespread use.

Manufacturers of wafer-based cells have responded to high silicon prices in 2004–2008 prices with rapid reductions in silicon consumption. In 2008, according to Jef Poortmans, director of IMEC's organic and solar department, current cells use between eight and nine grams of silicon per watt of power generation, with wafer thicknesses in the neighbourhood of 0.200 mm.

Current Events

Other technologies have tried to enter the market. First Solar was briefly the largest panel manufacturer in 2009, in terms of yearly power produced, using a thin-film cell sandwiched between two layers of glass. Since then silicon panels reasserted their dominant position both in terms of lower prices and the rapid rise of Chinese manufacturing, resulting in the top producers being Chinese. By late 2011, efficient production in China, coupled with a drop in European demand due to budgetary turmoil had dropped prices for crystalline solar-based modules further, to about $1.09 per watt in October 2011, down sharply from the price per watt in 2010. Prices continued to fall in 2012, reaching $0.62/watt by 4Q2012 with further price reductions expected in 2013.

A more modern process, mono-like-multi, aims to offer the performance of mono at the cost of poly, and is in the process of being introduced in 2012.

As of 2013, the plummeting cost of silicon still continues to usher in a new era of inexpensive solar cells, making non-cell and even non-module costs ever more important for the advancement of economical solar energy.

Applications

Figure: *Polycrystalline photovoltaic cells laminated to backing material in a module*

Figure: *Polycrystalline photovoltaic cells*

Solar cells are often electrically connected and encapsulated as a module. Photovoltaic modules often have a sheet of glass on the front (sun up) side, allowing light to pass while protecting the semiconductor wafers from abrasion and impact due to wind-driven debris, rain, hail, etc. Solar cells are also usually connected in series in modules, creating an additive voltage. Connecting cells in parallel will yield a higher current; however, very significant problems exist with parallel connections. For example, shadow effects can shut down the weaker (less illuminated) parallel string (a number of series connected cells) causing substantial power loss and even damaging the weaker string because of the excessive reverse bias applied to the shadowed cells by their illuminated partners. Strings of series cells are usually handled independently and not connected in parallel, special paralleling circuits are the exceptions. Although modules can be interconnected to create an array with the desired peak DC voltage and loading current capacity, using independent MPPTs (maximum power point trackers) provides a better solution. In the absence of paralleling circuits, shunt diodes can be used to reduce the power loss due to shadowing in arrays with series/parallel connected cells.

To make practical use of the solar-generated energy, the electricity is most often fed into the electricity grid using inverters (grid-connected photovoltaic systems); in stand-alone systems, batteries are used to store the energy that is not needed immediately. Solar panels can be used to power or recharge portable devices.

Theory

The solar cell works in three steps:

1. Photons in sunlight hit the solar panel and are absorbed by semiconducting materials, such as silicon.
2. Electrons (negatively charged) are excited from their current molecular/atomic orbital. Once excited the electron can either dissipate the energy, and return to its orbital or travel through the cell until it reaches an electrode. Current starts flowing through the material to cancel the potential and this electricity is captured. Due to the special composition of solar cells, the electrons are only allowed to move in a single direction.
3. An array of solar cells converts solar energy into a usable amount of direct current (DC) electricity.

Efficiency

The efficiency of a solar cell may be broken down into reflectance efficiency, thermodynamic efficiency, charge carrier separation efficiency and conductive efficiency. The overall efficiency is the product of all of these individual efficiencies.

A solar cell usually has a voltage dependent efficiency curve, temperature coefficients, and shadow angles.

Due to the difficulty in measuring these parameters directly, other parameters are measured instead: thermodynamic efficiency, quantum efficiency, integrated quantum efficiency, V_{OC} ratio, and fill factor. Reflectance losses are a portion of the quantum efficiency under "external quantum efficiency". Recombination losses make up a portion of the quantum efficiency, V_{OC} ratio, and fill factor.

Resistive losses are predominantly categorized under fill factor, but also make up minor portions of the quantum efficiency, V_{OC} ratio.

The fill factor is defined as the ratio of the actual maximum obtainable power to the product of the open circuit voltage and short circuit current. This is a key parameter in evaluating the performance of solar cells. Typical commercial solar cells have a fill factor > 0.70. Grade B cells have a fill factor usually between 0.4 to 0.7. Cells with a high fill factor have a low equivalent series resistance and a high equivalent shunt resistance, so less of the current produced by the cell is dissipated in internal losses.

Single p–n junction crystalline silicon devices are now approaching the theoretical limiting power efficiency of 33.7%, noted as the Shockley–Queisser limit in 1961. In the extreme, with an infinite number of layers, the corresponding limit is 86% using concentrated sunlight.

Figure: *Solar panels on the International Space Station absorb light from both sides. These Bifacial cells are more efficient and operate at lower temperature than single sided equivalents.*

Other Ways for Improving the Efficiency

There are many reviews that suggest many ways to improve the solar cells efficiency, consisting to incorporate lanthanide doped materials (Er^{3+}, Yb^{3+}, Ho^{3+} or combination of them) on to a solar cell, taking advantage of their optical properties to convert the infrared radiation to visible light . This kind of process, called upconversion, is a luminescence process where two low-energy infrared photons are absorbed by rare-earth ions to generate a high-energy visible photon. As example, the energy transfer upconversion process (ETU), consists in successive transfer processes between excited ions under excitation in the near infrared. Therefore, an excited ion can reach a high energetic level from which de-excites emitting a visible photon. By this way, an upconverter material could be placed below the solar cell in order to absorb the infrared light (not absorbed by the Si) and produce visible photons to excite the solar cell, enhancing its response to infrared radiation. There are many upconversion materials, but usually they are doped with lanthanide ions because they are efficient in these upconversion processes. These ions are most commonly found in the trivalent state and their rich energy level structure over a broad spectral range explains their numerous applications. Among them, the Er^{3+} ions have been the most used. The Er^{3+} ions absorb the solar radiation around 1.54 μm. Two Er^{3+} ions that have absorbed this radiation can interact with each other through an upconversion process. The excited ion emits light above the Si bandgap that is absorbed by the solar cell and creates an additional electron–hole pair that can contribute to generate current. However, the obtained efficiency of the solar cell was small. In addition, the fluoroindate glasses have shown to have very low phonon energy and they have been also proposed as suitable matrix doped with Ho^{3+} ions in order to improve the efficiency of silicon solar cells, using upconversion processes.

Cost

The cost of a solar cell is given per unit of peak electrical power. Solar-specific feed-in tariffs vary worldwide, and even state by state within various countries. Such feed-in tariffs can be highly effective in encouraging the development of solar power projects.

High-efficiency solar cells are of interest to decrease the cost of solar energy. Many of the costs of a solar power plant are proportional to the panel area or land area of the plant. A higher efficiency cell may reduce the required areas and so reduce the total plant cost, even if the cells themselves are more costly. Efficiencies of bare cells, to be useful in evaluating solar power plant economics, must be evaluated under realistic conditions. The basic parameters that need to be evaluated are the short circuit current, open circuit voltage.

The chart above illustrates the best laboratory efficiencies obtained for various materials and technologies, generally this is done on very small, i.e., one square cm, cells. Commercial efficiencies are significantly lower.

Grid parity, the point at which photovoltaic electricity is equal to or cheaper than grid power, can be reached using low cost solar cells. Proponents of solar hope to achieve grid parity first in areas with abundant sun and high costs for electricity such as in California and Japan. Some argue that grid parity has been reached in Hawaii and other islands that otherwise use diesel fuel to produce electricity. George W. Bush had set 2015 as the date for grid parity in the USA. Speaking at a conference in 2007, General Electric's Chief Engineer predicted grid parity without subsidies in sunny parts of the United States by around 2015.

The Photovoltaic Association reported in 2012 that Australia has reached grid parity (ignoring feed in tariffs).

The price of solar panels fell steadily for 40 years, until 2004 when high subsidies in Germany drastically increased demand there and greatly increased the price of purified silicon (which is used in computer chips as well as solar panels). The recession of 2008 and the onset of Chinese manufacturing caused prices to resume their decline with vehemence. In the four years after January 2008 prices for solar modules in Germany dropped from €3 to €1 per peak watt. During that same times production capacity surged with an annual growth of more than 50%. China increased market share from 8% in 2008 to over 55% in the last quarter of 2010. Recently, in December 2012 the price of Chinese solar panels had dropped to \$0.60/Wp (crystalline modules).

Swanson's law, an observation similar to Moore's Law that states that solar cell prices fall 20% for every doubling of industry capacity, has gained recent (as of 2012) media attention, having been featured in an article in the British weekly newspaper The Economist.

Practical Materials

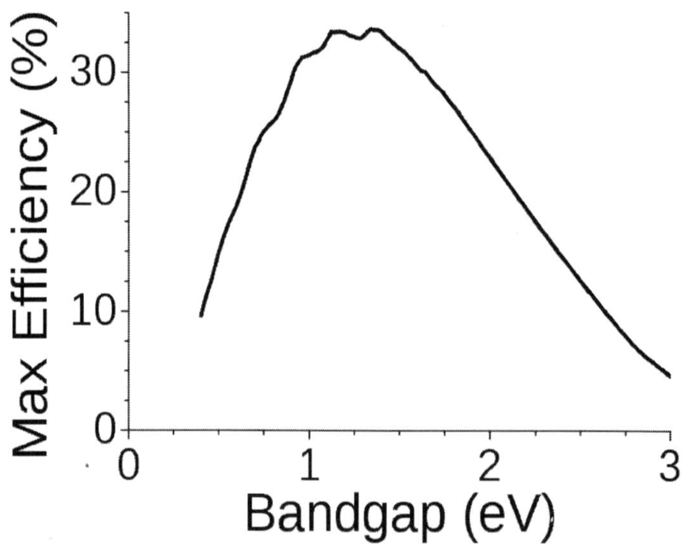

Figure: *The Shockley-Queisser limit for the theoretical maximum efficiency of a solar cell. Semiconductors with band gap between 1 and 1.5eV, or near-infrared light, have the greatest potential to form an efficient single-junction cell. (The efficiency "limit" shown here can be exceeded by multijunction solar cells.)*

Various materials display varying efficiencies and have varying costs. Materials for efficient solar cells must have characteristics matched to the spectrum of available light. Some cells are designed to efficiently convert wavelengths of solar light that reach the Earth surface. However, some solar cells are optimized for light absorption beyond Earth's atmosphere as well. Light absorbing materials can often be used in *multiple physical configurations* to take advantage of different light absorption and charge separation mechanisms.

Industrial photovoltaic solar cells are made of monocrystalline silicon, polycrystalline silicon, amorphous silicon, cadmium telluride or copper indium selenide/sulfide, or GaAs-based multijunction material systems.

Many currently available solar cells are made from bulk materials that are cut into wafers between 180 to 240 micrometres thick that are then processed like other semiconductors.

Crystalline Silicon

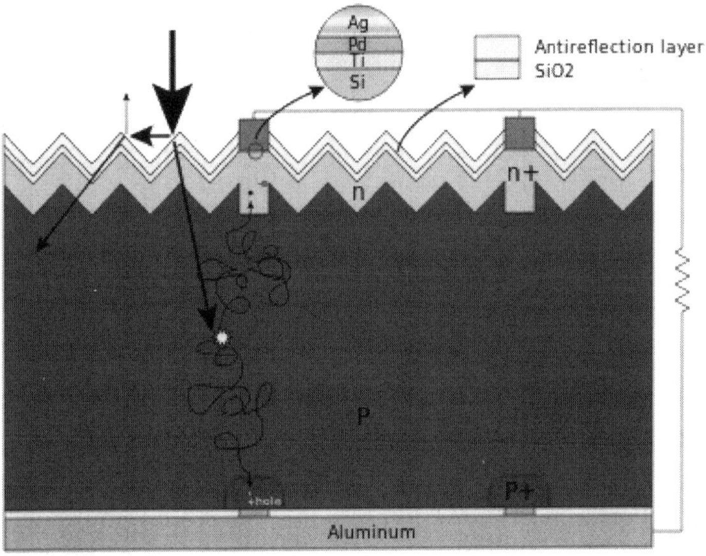

Figure: *Basic structure of a silicon based solar cell and its working mechanism.*

By far, the most prevalent *bulk* material for solar cells is crystalline silicon (abbreviated as a group as *c-Si*), also known as "solar grade silicon". Bulk silicon is separated into multiple categories according to crystallinity and crystal size in the resulting ingot, ribbon, or wafer. These cells are entirely based around the concept of a p-n junction.

1. *monocrystalline silicon* (c-Si): often made using the Czochralski process. Single-crystal wafer cells tend to be expensive, and because they are cut from cylindrical ingots, do not completely cover a square solar cell module without a substantial waste of refined silicon. Hence most *c-Si* panels have uncovered gaps at the four corners of the cells.
2. *polycrystalline silicon*, or *multicrystalline silicon*, (poly-Si or mc-Si): made from cast square ingots — large blocks of molten silicon carefully

cooled and solidified. Poly-Si cells are less expensive to produce than single crystal silicon cells, but are less efficient. United States Department of Energy data show that there were a higher number of polycrystalline sales than monocrystalline silicon sales.

3. *ribbon silicon* is a type of polycrystalline silicon: it is formed by drawing flat thin films from molten silicon and results in a polycrystalline structure. These cells have lower efficiencies than poly-Si, but save on production costs due to a great reduction in silicon waste, as this approach does not require sawing from ingots.
4. *mono-like-multi silicon*: Developed in the 2000s and introduced commercially around 2009, mono-like-multi, or cast-mono, uses existing polycrystalline casting chambers with small "seeds" of mono material. The result is a bulk mono-like material with poly around the outsides. When sawn apart for processing, the inner sections are high-efficiency mono-like cells (but square instead of "clipped"), while the outer edges are sold off as conventional poly. The result is line that produces mono-like cells at poly-like prices.

Thin Films

Thin-film technologies reduce the amount of material required in creating the active material of solar cell. Most thin film solar cells are sandwiched between two panes of glass to make a module. Since silicon solar panels only use one pane of glass, thin film panels are approximately twice as heavy as crystalline silicon panels, although they have a smaller ecological impact (determined from life cycle analysis). The majority of film panels have significantly lower conversion efficiencies, lagging silicon by two to three percentage points. Thin-film solar technologies have enjoyed large investment due to the success of First Solar and the largely unfulfilled promise of lower cost and flexibility compared to wafer silicon cells, but they have not become mainstream solar products due to their lower efficiency and corresponding larger area consumption per watt production. Cadmium telluride (CdTe), copper indium gallium selenide (CIGS) and amorphous silicon (a-Si) are three thin-film technologies often used as outdoor photovoltaic solar power production. As of December 2013, CdTe was most cost effective (U.S. manufacturing cost per installed watt: $0.59 reported by First Solar) widely used thin film technology, and CIGS technology has the highest laboratory efficiency (20.4% as of December 2013), though CdTe cells made by First Solar have the highest industrial efficiency, and the lab efficiency of the immature GaAs thin film technology tops 28%.

Cadmium Telluride Solar Cell

Cadmium telluride (CdTe) photovoltaics describes a photovoltaic (PV) technology that is based on the use of cadmium telluride thin film, a semiconductor layer designed to absorb and convert sunlight into electricity. Cadmium telluride PV is the only thin film photovoltaic technology to surpass crystalline silicon PV in cheapness for a significant portion of the PV market, namely in multi-kilowatt systems. CdTe PV is considered the ecologically leading technology as it provides a solution to key ecological issues including climate change, energy security, and water scarcity. It is also considered the most eco-efficient current PV technology when comparing a range of application scenarios e.g. commercial rooftop applications or large scale ground mount applications. On a life cycle basis, CdTe PV has the smallest carbon footprint, lowest water use, and fastest energy payback time of all solar technologies. A fast energy payback time enables PV to scale with faster carbon reductions without causing short-term energy deficits.

A cadmium telluride solar cell uses a cadmium telluride (CdTe) thin film, a semiconductor layer to absorb and convert sunlight into electricity. One disadvantage of this technology, the only thin film material so far to rival crystalline silicon in cost/watt, is that cadmium is a deadly poison. Another issue is that tellurium (anion: "telluride") is a metal extremely rare in the earth's crust. The cadmium present in the cells would be toxic if released. However, release is impossible during normal operation of the cells and is unlikely during fires in residential roofs. A square metre of CdTe contains approximately the same amount of Cd as a single C cell nickel-cadmium battery, in a more stable and less soluble form.

Copper Indium Gallium Selenide

Copper indium gallium selenide ($CuIn_xGa_{1-x}Se_2$ or CIGS) is a direct bandgap semiconductor useful for the manufacture of solar cells. The CIGS absorber is deposited on a glass or plastic backing, along with electrodes on the front and back to collect current. Because the material has a high absorption coefficient and strongly absorbs sunlight, a much thinner film is required than of other semiconductor materials. Devices made with CIGS belong to the thin-film category of photovoltaics (PV).

CIGS is one of three mainstream thin-film PV technologies, the other two being cadmium telluride and amorphous silicon. Like these materials, CIGS layers are thin enough to be flexible, allowing them to be deposited on flexible substrates. However, as all of these technologies normally use high-temperature deposition techniques, the best performance normally comes from cells deposited on glass. Even then the performance is marginal compared to modern polysilicon-based panels. Advances in low-temperature deposition of CIGS cells has erased much of this performance difference, even with flexible designs.

The market for thin-film PV grew at a 60% annual rate from 2002 to 2007 and is still growing rapidly. Therefore, a strong incentive exists to develop and improve deposition methods for these films that will allow lower cost and increased throughput.

Properties

CIGS is a I-III-VI_2 compound semiconductor material composed of copper, indium, gallium, and selenium. The material is a solid solution of copper indium selenide (often abbreviated "CIS") and copper gallium selenide, with a chemical formula of $CuIn_xGa_{(1-x)}Se_2$, where the value of x can vary from 1 (pure copper indium selenide) to 0 (pure copper gallium selenide). It is a tetrahedrally bonded semiconductor, with the chalcopyrite crystal structure. The bandgap varies continuously with *x* from about 1.0 eV (for copper indium selenide) to about 1.7 eV (for copper gallium selenide).

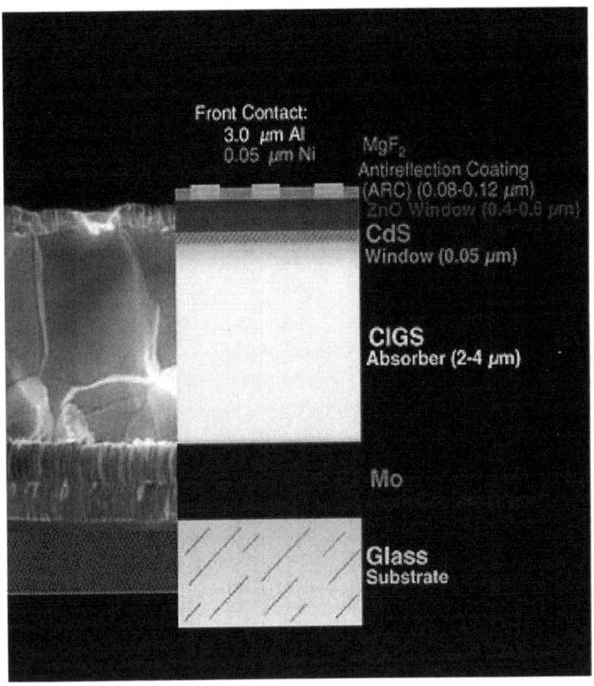

Figure: *CIGS device structure*

CIGS has an exceptionally high absorption coefficient of more than 10^5/cm for 1.5 eV and higher energy photons. CIGS solar cells with efficiencies around 20% have been claimed by both the National Renewable Energy Laboratory (NREL) and the Zentrum für Sonnenenergie und Wasserstoff Forschung (ZSW), which is the record to date for any thin film solar cell.

CIGS Photovoltaic Cells

The most common device structure for CIGS solar cells is shown in Figure. Soda lime glass is commonly used as a substrate, because it contains Na, which has been shown to yield a substantial open-circuit voltage increase, notably through surface and grain boundary defects passivation. However, many companies are also looking at lighter and more flexible substrates such as polyimide or metal foils. A molybdenum layer is deposited (commonly by sputtering) which serves as the back contact and to reflect most unabsorbed light back into the absorber. Following Mo deposition a p-type CIGS absorber layer is grown by one of several unique methods. A thin n-type buffer layer is added on top of the absorber. The buffer is typically CdS deposited via chemical bath deposition. The buffer is overlaid with a thin, intrinsic ZnO layer (i-ZnO) which is capped by a thicker, Al doped ZnO layer. The i-ZnO layer is used to protect the CdS and the absorber layer from sputtering damage while depositing the ZnO:Al window layer, since the latter is usually deposited by DC sputtering, known as a damaging process. The Al doped ZnO serves as a transparent conducting oxide to collect and move electrons out of the cell while absorbing as little light as possible.

The materials based on $CuInSe_2$ that are of interest for photovoltaic applications include several elements from groups I, III and VI in the periodic table. These semiconductors are especially attractive for thin film solar cell application because of their high optical absorption coefficients and versatile optical and electrical characteristics which can in principle be manipulated and tuned for a specific need in a given device.

Conversion Efficiency

CIGS is mainly used in the form of polycrystalline thin films. The best efficiency achieved as of October 2013 was 20.8%. A team at the National Renewable Energy Laboratory achieved 19.9% new world record efficiency by modifying the CIGS surface and making it look like CIS. All of these high-performance examples were deposited on glass, which meant the products were not flexible. In 2013, scientists at Empa, the Swiss Federal Laboratories for Materials Science and Technology, developed CIGS cells on flexible polymer foils with a new record efficiency of 20.4% for converting sunlight into electricity. These are both the highest efficiency examples as well as flexible.

These efficiencies are different from module conversion efficiencies. Two of the leading manufacturers of CIGS thin-film PV have hit new record highs in module conversion efficiencies. The U.S. National Renewable Energy Laboratory has confirmed 13.8% efficiency of a large-area (metre-square) production panel, and 13% total-area (and 14.2% aperture-area) efficiency with some production modules. In September 2012 the German Manz AG presented a CIGS solar module with an efficiency of 14.6 % on total module surface and 15.9 % on aperture, which was produced on a mass production facility. MiaSolé obtained a certified 15.7% aperture-area efficiency on a $1 m^2$ production module, and Solar Frontier claimed a 17.8% efficiency on a $900 cm^2$ module.

Higher efficiencies (around 30%) can be obtained by using optics to concentrate the incident light. The use of gallium increases the optical band gap of the CIGS layer as compared to pure CIS, thus increasing the open-circuit voltage. In another point of view, gallium is added to replace as much indium as possible due to gallium's relative availability to indium.

Deposition

CIGS films can be manufactured by several different methods:

- The most common vacuum-based process is to co-evaporate or co-sputter copper, gallium, and indium onto a substrate at room temperature, then anneal the resulting film with a selenide vapor to form the final CIGS structure. An alternative process is to co-evaporate copper, gallium, indium and selenium onto a heated substrate.
- A non-vacuum-based alternative process deposits nanoparticles of the precursor materials on the substrate and then sinters them in situ. Electroplating is another low cost alternative to apply the CIGS layer.

With record CIGS efficiency at just below 20% for several years, new trends of CIGS research has been focused on lower-cost deposition methods as an alternative to expensive vacuum processes. This new research progressed quickly and efficiencies of 10%–15% have been achieved by many teams.

CIGS and Silicon

Unlike the silicon cells based on a homojunction, the structure of CIGS cells is a more complex heterojunction system. CIGS solar cells are not as efficient as crystalline silicon solar cells, for which the record efficiency lies at 24.7%. Though some argue that CIGS will be substantially cheaper due to much lower material costs and potentially lower fabrication costs, in the second quarter of 2011, Veeco, Inc., a company with expertise in thin film deposition technology, closed a major CIGS development project due in part to the continued declining costs of the competing silicon materials and technology.

A direct bandgap material, CIGS has very strong light absorption and only 1–2 micrometres of CIGS is enough to absorb most of the sunlight. A much greater thickness of crystalline silicon is required for the same absorption.

The active layer (CIGS) can be deposited in a polycrystalline form directly onto molybdenum coated glass sheets or steel bands. This uses less energy than growing large crystals, which is a necessary step in the manufacture of crystalline silicon solar cells. Also unlike crystalline silicon, these substrates can be flexible.

CIGS and Other Thin Films

CIGS can be used to make thin film solar cells (TFSC). "Thinness" is somewhat of a relative term, but in photovoltaics it generally is in reference to so-called "first generation" high-efficiency silicon cells, which are manufactured from bulk wafers hundreds of micrometres thick. CIGS has much higher absorption than silicon, so a layer of CIGS can absorb more light than a silicon layer of the same thickness. With thin films, some of the light gathering efficiency is given up in exchange for the advantages of thinness, mainly the use of less material in manufacturing, and in highly absorptive CIGS the efficiency tradeoff is less severe than in silicon. The record efficiencies for thin film CIGS cells are slightly lower than that of CIGS for lab-scale top performance cells. The 19.9% efficiency is by far the highest compared with those achieved by other thin film technologies such as Cadmium Telluride (CdTe) or amorphous silicon (a-Si). As for CIS and CGS solar cells, the world record total area efficiencies are 15.0% and 9.5% respectively. Another advantage of CIGS compared to CdTe is smaller amount of toxic material cadmium are present in CIGS cells.

Structure of a CIGS Thin-Film Solar Cell

Figure: *Cross-section of an $Cu(In,Ga)Se_2$ solar cell*

The basic structure of a $Cu(In,Ga)Se_2$ thin-film solar cell is depicted in the image of Figure to the right. The most common substrate is soda-lime glass of 1–3 mm thickness. This is coated on one side with molybdenum (Mo) that serves as metal back contact. The heterojunction is formed between the semiconductors CIGS and ZnO, separated by a thin layer of CdS and a layer of intrinsic ZnO. The CIGS is doped p-type from intrinsic defects, while the ZnO is doped n-type to a much larger extent through the incorporation of aluminum (Al). This asymmetric doping causes the space-charge region to extend much further into the CIGS than into the ZnO. Matched to this are the layer thicknesses and the bandgaps of the materials: the wide CIGS layer serves as absorber with a bandgap between 1.02 eV ($CuInSe_2$) and 1.65 eV ($CuGaSe_2$). Absorption is minimized in the upper layers, called window, by the choice of larger bandgaps: $E_{g,ZnO}$=3.2 eV and $E_{g,CdS}$=2.4 eV. The doped ZnO also serves as front contact for current collection. Laboratory scale devices, typically 0.5 cm^2 in size, are provided with a Ni/Al-grid deposited onto the front side to contact the ZnO.

Production of modules involves the deposit on layer being cut into a series of parallel connected strips. A further transparent protective cover is applied to the module. This sandwich construction is then sealed against the ingress of moisture. Some method of physical support is required to prevent fracture of this fragile structure.

General Properties of High Performance CIGS Absorbers

All high performance CIGS absorbers in solar cells have several similarities independent of the growth technique used. First, they are polycrystalline α-phase which has the chalcopyrite crystal structure shown in Figure earlier. The second property is an overall Cu deficiency. Cu deficiency increases the majority carrier (hole) concentration by increasing the number of Cu vacancies. These vacancies act as electron acceptors.

Also, when CIGS films are In rich (Cu deficient) the surface layer of the film forms an ordered defect compound (ODC) with a stoichiometry of $Cu(In,Ga)_3Se_5$. The ODC is n-type, forming a p-n homojunction in the film at the interface between the α phase and the ODC. Recombination velocity at the CIGS/CdS interface is decreased by presence of the homojunction. The drop in interface recombination attributable to ODC formation is demonstrated by experiments which have shown recombination in the bulk of the film is the main loss mechanism in Cu deficient films, while in Cu rich films the main loss is at the CIGS/CdS interface.

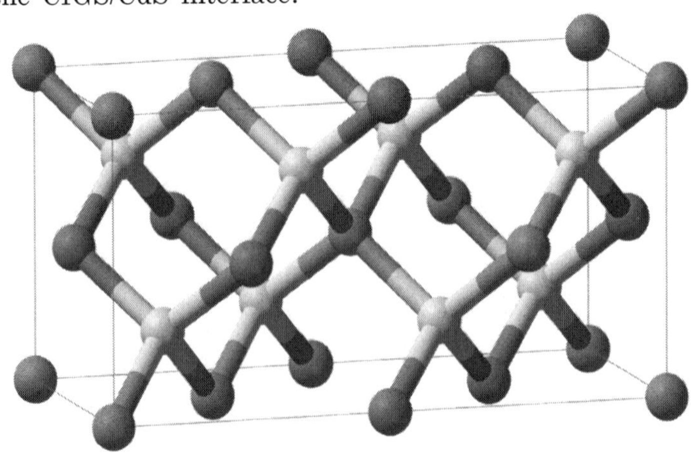

Figure: *CIGS unit cell. Red = Cu, Yellow = Se, Blue = In/Ga*

Sodium (Na) incorporation is also necessary for optimal performance. Ideal Na concentration is considered to be approximately 0.1 at%. Na is commonly supplied by the soda-lime glass used as the substrate, but in processes that do not use this substrate the Na must be deliberately added. Beneficial effects of Na include increases in p-type conductivity, texture, and average grain size.

Furthermore, Na incorporation allows for performance to be maintained over larger stoichiometric deviations. Simulations have predicted that Na on an In site creates a shallow acceptor level and that Na serves to remove In on Cu defects (donors), but reasons for these benefits are still being debated. Na is also credited with catalyzing oxygen absorption. Oxygen passivates Se vacancies that act as compensating donors and recombination centres.

Alloying CIS ($CuInSe_2$) with CGS ($CuGaSe_2$) increases in the bandgap. To reach the ideal bandgap for a single junction solar cell, 1.5 eV, a Ga/(In+Ga) ratio of roughly 0.7 would be optimal. However, at ratios above ~0.3 device performance drops off. Industry currently targets the 0.3 Ga/(In+Ga) ratio, resulting in bandgaps between 1.1 and 1.2 eV. The decreasing performance has been postulated to be a result of CGS not forming the ODC, which is necessary for a good interface with CdS.

The highest efficiency devices show a high degree of texturing, or preferred crystallographic orientation. Until recently record efficiency devices displayed a (112) texture, but now a (204) surface orientation is observed in the best quality devices. A smooth absorber surface is preferred to maximize the ratio of the illuminated area to the area of the interface. The area of the interface increases with roughness while illuminated area remains constant, decreasing open circuit voltage (V_{OC}). Studies have also linked an increase in defect density to decreased V_{OC}. Recombination in CIGS has been suggested to be dominated by non-radiative processes. Theoretically, recombination can be controlled by engineering of the film, as opposed to being intrinsic to the material.

Precursor Deposition and Post Processing

Perhaps the most common method used to create CIGS films for commercial use is deposition of precursor materials – always including Cu, In, and Ga, and sometimes including Se – onto a substrate and processing these films at high temperatures under a proper atmosphere. The following sections outline the various techniques for precursor deposition processing, including sputtering of metallic layers at low temperatures, printing of inks containing nanoparticles, electrodeposition, and a technique inspired by wafer-bonding.

General Selenization Concerns

The Se supply and selenization environment is extremely important in determining the properties and quality of the film produced from precursor layers. When Se is supplied in the gas phase (for example as H_2Se or elemental Se) at high temperatures the Se will become incorporated into the film by absorption and subsequent diffusion. During this step, called chalcogenization, complex interactions occur to form a chalcogenide. These interactions include formation of Cu-In-Ga intermetallic alloys, formation of intermediate metal-selenide binary compounds, and phase separation of various stoichiometric CIGS compounds. Because of the variety and complexity of the reactions taking place, the properties of the CIGS film are difficult to control.

Differences exist between films formed using different Se sources. Using H_2Se yields the fastest Se incorporation into the absorber; 50 at% Se can be achieved in CIGS films at temperatures as low as 400 °C. By comparison, elemental Se only achieves full incorporation with reaction temperatures of 500 °C and above. Below 500 °C films formed from elemental Se were not only Se deficient, but also had multiple phases including metal selenides and various alloys. Use of H_2Se also provides the best compositional uniformity and the largest grain sizes. However, H_2Se is highly toxic and is classified as hazardous to the environment.

Sputtering of Metallic Layers Followed by Selenization

In this method of forming CIGS absorbers, a metal film of Cu, In, and Ga is sputtered at or near room temperature and reacted in a Se atmosphere at high temperature. This process has higher throughput than coevaporation and compositional uniformity can be more easily achieved.

Sputtering a stacked multilayer of metal – for example a Cu/In/Ga/Cu/In/Ga... structure – produces a smoother surface and better crystallinity in the absorber, when compared to a simple bilayer (Cu-Ga alloy/In) or trilayer (Cu/In/Ga) sputtering. These attributes result in higher efficiency devices, but forming the multilayer is a more complicated deposition process and is likely not worth the cost of extra equipment or the added process complexity. Additionally, the reaction rates of Cu/Ga and Cu/In layers with Se are different. If the reaction temperature is not high enough, or not held long enough, CIS and CGS form as separate phases. The same considerations outlined in the previous section apply to Se incorporation.

Companies currently using similar processes include Showa Shell, Avancis (now an affiliate of Saint-Gobain Group), Miasolé, Honda Soltec, and Energy Photovoltaics (EPV). Showa Shell sputters a Cu-Ga alloy layer and an In layer, followed by selenization in H_2Se and sulfurization in H_2S. The sulfurization step appears to passivate the surface in a way similar to CdS in most other cells. Thus, the buffer layer used is Cd-free which eliminates the worries related to the toxicity and environmental impact of Cd. Showa Shell has reported a maximum module efficiency of 13.6% with an average of 11.3% for 3600 cm^2 substrates. Shell Solar uses the same technique as Showa Shell to create the absorber; however, they use a CdS layer deposited by chemical vapor deposition. Modules sold by Shell Solar have a specification of 9.4% module efficiency.

Miasole has had great success in procuring venture capital funds for its process and scale up. However, little is known about their sputtering/selenization process beyond their stated efficiency of 9 to 10% for modules.

EPV uses a hybrid between coevaporation and sputtering in which In and Ga are evaporated in a Se atmosphere. This is followed by Cu sputtering and a selenization step. Finally, In and Ga are again evaporated in the presence of Se. Based on Hall measurements, these films have a low carrier concentration and high mobility compared to other devices. EPV films have also been shown to have a low defect concentration.

Chalcogenization of Particulate Precursor Layers

In this method, metal or metal-oxide nanoparticles are used as the precursors for CIGS growth. These nanoparticles are generally suspended in a water based solution and then applied to large areas by various methods, with printing the most common. The film is then dehydrated and, if the precursors are metal-oxides, reduced in a H_2/N_2 atmosphere. Following dehydration, the remaining porous film is sintered and selenized at temperatures greater than 400 °C.

Nanosolar and International Solar Electric Technology (ISET) are attempting to scale up this process. ISET uses oxide particles while Nanosolar is extremely secretive about their ink. The ink composition is unknown but there is some implication that Se is also incorporated into Nanosolar's ink. The advantages of this process include uniformity over large areas, non-vacuum or low-vacuum equipment, and adaptability to roll-to-roll manufacturing. When compared to laminar metal precursor layers, the selenization of sintered nanoparticles is more rapid. The increased rate is a result of the greater surface area associated with porosity. Decreasing high temperature selenization reduces the thermal budget. Unfortunately, the drawback of porosity is a tendency towards rougher absorber surfaces. Use of particulate precursors allows for printing on a large variety of substrates with high materials utilisation, around 90% or more. A disadvantage is that little research and development exists in this area of deposition. In Nanosolar's manufacturing the printed rolls are cut into cells and must be binned and integrated in a fashion similar to how Si devices are made today. The binning process is different from the monolithic integration that many CIGS companies are using. Monolithic integration is far more adaptable to inline production.

Nanosolar has reported a cell (not module) efficiency of 14%, however this has not been verified by any national laboratory testing, nor are they allowing any onsite inspections of their facilities to verify this and other claims made in the past. In independent testing, ISET's absorber had the 2nd lowest efficiency at 8.6%. However, all the modules that beat ISET's module were coevaporated, a process which has manufacturing disadvantages and higher costs. ISET's sample suffered most from low V_{OC} and low fill factor, indicative of a rough surface and/or a high number of defects aiding recombination. Related to these issues, the film had poor transport properties including a low Hall mobility and short carrier lifetime.

Electrodeposition Followed by Selenization

Precursors can also be deposited by electrodeposition. Two different methodologies exist: deposition of elemental layered structures, and simultaneous deposition of all elements (including Se). Both methods require thermal treatment in a Se atmosphere to make device quality films. Because electrodeposition requires conductive electrodes, metal foils are a logical substrate. Electrodeposition of elemental layers is similar to the sputtering of elemental layers. Currently no company is scaling up this process.

Simultaneous deposition is performed using a working electrode (cathode), a counter electrode (anode), and a reference electrode as in Figure. A metal foil substrate is used as the working electrode in industrial processes. An inert material is used for the counter electrode, and the reference electrode exists to measure and control the potential difference between the anode and cathode. The reference electrode allows the process to be performed potentiostatically, meaning the potential of the substrate can be controlled.

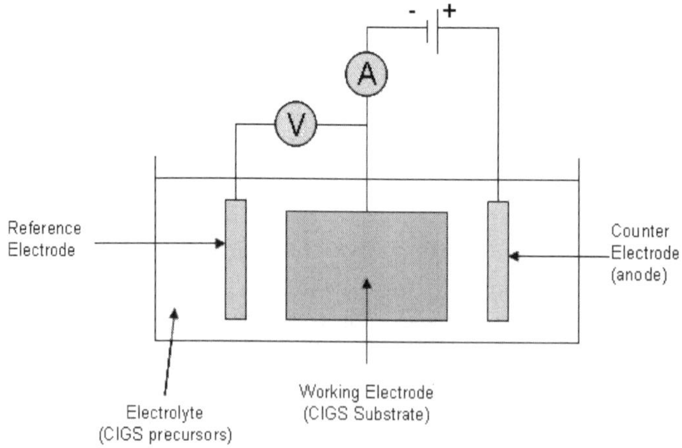

Figure: *CIGS electrodeposition apparatus*

Electrodeposition of all elements simultaneously is a difficult processing problem for a variety of reasons. First, the standard reduction potentials of the elements are not the same, causing preferential deposition of a single element. This problem is commonly alleviated by adding different counter ions into solution for each ion to be deposited (Cu^{2+}, Se^{4+}, In^{3+}, and Ga^{3+}), thus changing the reduction potential for that ion. Second, the Cu-Se system has a complicated behaviour and the composition of the film depends on the Se^{4+}/Cu^{2+} ion flux ratio which can vary over the film surface. Because of this behaviour the deposition conditions, specifically precursor concentrations and deposition potential, need to be optimized. Even with optimization, reproducibility is low over large areas due to composition variations and potential drops along the substrate.

The resulting films have small grains, are Cu-rich, and generally contain $Cu_{2-x}Se_x$ phases along with impurities from the solution. Annealing is required to improve crystallinity. In order to achieve efficiencies higher than 7%, a stoichiometry correction is also required.

The correction is done via high temperature physical vapor deposition which is not practical in industry.

Having solved the optimization issues, Solopower is currently producing cells with >13.7% conversion efficiency as per NREL. SoloPower is currently attempting to scale up the process, but few details have been released. SoloPower is relying on the advantages of roll-to-roll manufacturing and flexible metal foil substrates.

Precursor Combination by Wafer-Bonding Inspired Technique

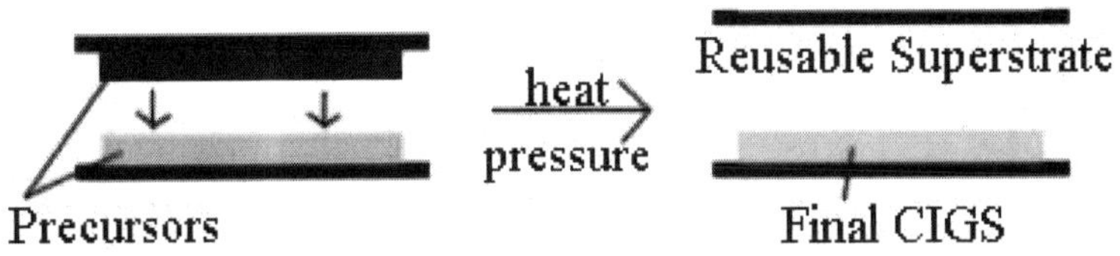

Figure: *Schematic of wafer-bonding inspired technique*

In this process, two different precursor films are deposited separately on a substrate and a superstrate. The films are pressed together and heated to release the film from the superstrate, leaving a CIGS absorber on the substrate. This technique allows the superstrate to be reused. Heliovolt patented this procedure and has named it the FASST process. Therefore, Heliovolt is the only company currently scaling up the technique. In principle, the precursors can be deposited at low temperature using low-cost deposition techniques, lowering the final cost of the module. However, the first one or two generations of product will still use higher temperature PVD methods and not achieve full cost cutting potential. Flexible substrates could eventually be used in this process.

Typical film characteristics are not known outside of the company, as no research has been conducted by independently funded laboratories. However, Heliovolt has claimed a top cell efficiency of 12.2%.

Coevaporation

Coevaporation, or codeposition, is the most prevalent CIGS fabrication technique in the laboratory and an important method in industry. The Boeing coevaporation process deposits bilayers of CIGS with different stoichiometries onto a heated substrate and allows them to intermix. The National Renewable Energy Laboratory (NREL) has developed another process which involves three deposition steps and produced the current CIGS efficiency record holder at 20.3%. The first step in NREL's method is codeposition of In, Ga, and Se. This is followed by Cu and Se deposited at a higher temperature to allow for diffusion and intermixing of the elements.

In the final stage In, Ga, and Se are again deposited to make the overall composition Cu deficient.

Würth Solar has been producing CIGS cells using an inline coevaporation system since 2005 with module efficiencies between 11% and 12% by the end of that year. They have subsequently opened another production facility and continued to improve efficiency and yield. Other companies scaling up coevaporation processes include Global Solar and Ascent Solar. Global Solar also uses an inline three stage deposition process. In all of the steps Se is supplied in excess in the vapor phase. In and Ga are first evaporated followed by Cu and then by In and Ga to make the film Cu deficient.

These films performed quite favourably in relation not only to other manufacturers but also to absorbers grown at NREL and the Institute for Energy Conversion (IEC). However, fully fabricated modules of Global Solar's films did not perform as well. The property in which the module most obviously under-performed was a low V_{OC}, which is characteristic of high defect density and high recombination velocities. Interestingly, Global Solar's absorber layer outperformed the NREL absorber in carrier lifetime and hall mobility. However, as completed cells the NREL sample performed better. This is evidence of a poor CIGS/CdS interface, possibly due to the lack of an ODC surface layer on the Global Solar film.

As most of the CIGS research at national laboratories and universities covers coevaporation, companies using this technique stand to gain the most from the scientific community. However, they also face significant disadvantages including uniformity issues over large areas and the related difficulty of coevaporating elements in an inline system. Another disadvantage is high growth temperatures which raise the thermal budget and cost. Additionally, coevaporation is plagued by low material utilisation (deposition on chamber walls instead of the substrate, especially for selenium) and expensive vacuum equipment. A way to enhance the utilisation of selenium is the usage of a thermal or plasma-enhanced selenium-cracking process, which can additionally be coupled with an ion beam source for ion beam assisted deposition.

Chemical Vapor Deposition

Chemical vapor deposition (CVD) has been implemented in multiple ways for the deposition of CIGS. Processes include atmosphere pressure metal organic CVD (AP-MOCVD), plasma-enhanced CVD (PECVD), low-pressure MOCVD (LP-MOCVD), and aerosol assisted MOCVD (AA-MOCVD). Current work is focused on changing the typical dual-source precursors to single-source precursors. Multiple source precursors must be homogeneously mixed and the flow rates of the precursors have to be kept at the proper stoichiometry. Single-source precursor methods do not suffer from these drawbacks and should enable better control of film composition compared to multiple source precursors.

CVD is not yet being used by any companies for CIGS synthesis. Currently, CVD produced films have low efficiency and a low V_{OC}, partially a result of a high defect concentration. Additionally, the surfaces of the films are generally quite rough which serves to further decrease the V_{OC}. However, the requisite Cu deficiency has been achieved using AA-MOCVD along with a (112) crystal orientation. However, if the film quality produced by CVD can be improved, any company using this technique could benefit from knowledge gained in other industries using large area CVD deposition, such as glass coating manufacturers. CVD deposition temperatures are lower than those used for other processes such as co-evaporation and selenization of metallic precursors. Therefore, CVD has a lower thermal budget, reducing the cost. Potential manufacturing problems include difficulties converting CVD to an inline process as well as the expense of handling volatile precursors.

GaAs Thin Film Cells

The Dutch Radboud University Nijmegen set the record for thin film solar cell efficiency using a single junction GaAs to 25.8% in August 2008 using only 4 μm thick GaAs layer which can be transferred from a wafer base to glass or plastic film. Recently, this record has been increased to 28.8%. The high efficiency obtained in GaAs thin film solar cells is attributed to the extreme high quality GaAs epitaxial growth, surface passivation by the AlGaAs, and the promotion of photon recycling by the thin film design.

Silicon Thin Films

Silicon thin-film cells are mainly deposited by chemical vapor deposition (typically plasma-enhanced, PE-CVD) from silane gas and hydrogen gas. Depending on the deposition parameters, this can yield:

1. Amorphous silicon (a-Si or a-Si:H)
2. Protocrystalline silicon or
3. Nanocrystalline silicon (nc-Si or nc-Si:H), also called microcrystalline silicon.

It has been found that protocrystalline silicon with a low volume fraction of nanocrystalline silicon is optimal for high open circuit voltage. These types of silicon present dangling and twisted bonds, which results in deep defects (energy levels in the bandgap) as well as deformation of the valence and conduction bands (band tails). The solar cells made from these materials tend to have lower *energy conversion efficiency* than *bulk* silicon, but are also less expensive to produce. The quantum efficiency of thin film solar cells is also lower due to reduced number of collected charge carriers per incident photon.

An amorphous silicon (a-Si) solar cell is made of amorphous or microcrystalline silicon and its basic electronic structure is the p-i-n junction. a-Si is attractive as a solar cell material because it is abundant and non-toxic (unlike its CdTe counterpart) and requires a low processing temperature, enabling production of devices to occur on flexible and low-cost substrates. As the amorphous structure has a higher absorption rate of light than crystalline cells, the complete light spectrum can be absorbed with a very thin layer of photo-electrically active material. A film only 1 micrometre thick can absorb 90% of the usable solar energy. This reduced material requirement along with current technologies being capable of large-area deposition of a-Si, the scalability of this type of cell is high. However, because it is amorphous, it has high inherent disorder and dangling bonds, making it a bad conductor for charge carriers. These dangling bonds act as recombination centres that severely reduce the carrier lifetime and pin the Fermi level so that doping the material to n- or p- type is not possible. Amorphous Silicon also suffers from the Staebler-Wronski effect, which results in the efficiency of devices utilising amorphous silicon dropping as the cell is exposed to light. The production of a-Si thin film solar cells uses glass as a substrate and deposits a very thin layer of silicon by plasma-enhanced chemical vapor deposition (PECVD). A-Si manufacturers are working towards lower costs per watt and higher conversion efficiency with continuous research and development on multijunction solar cells for solar panels. Anwell Technologies Limited recently announced its target for multi-substrate-multi-chamber PECVD, to lower the cost to US $0.50 per watt.

Amorphous silicon has a higher bandgap (1.7 eV) than crystalline silicon (c-Si) (1.1 eV), which means it absorbs the visible part of the solar spectrum more strongly than the infrared portion of the spectrum. As nc-Si has about the same bandgap as c-Si, the nc-Si and a-Si can advantageously be combined in thin layers, creating a layered cell called a tandem cell. The top cell in a-Si absorbs the visible light and leaves the infrared part of the spectrum for the bottom cell in nc-Si.

Recently, solutions to overcome the limitations of thin-film crystalline silicon have been developed. Light trapping schemes where the weakly absorbed long wavelength light is obliquely coupled into the silicon and traverses the film several times can significantly enhance the absorption of sunlight in the thin silicon films. Minimizing the top contact coverage of the cell surface is another method for reducing optical losses; this approach simply aims at reducing the area that is covered over the cell to allow for maximum light input into the cell. Anti-reflective coatings can also be applied to create destructive interference within the cell. This can be done by modulating the refractive index of the surface coating; if destructive interference is achieved, there will be no reflective wave and thus all light will be transmitted into the semiconductor cell. Surface texturing is another option, but may be less viable because it also increases the manufacturing price.

By applying a texture to the surface of the solar cell, the reflected light can be refracted into striking the surface again, thus reducing the overall light reflected out. Light trapping as another method allows for a decrease in overall thickness of the device; the path length that the light will travel is several times the actual device thickness. This can be achieved by adding a textured backreflector to the device as well as texturing the surface. If both front and rear surfaces of the device meet this criterion, the light will be 'trapped' by not having an immediate pathway out of the device due to internal reflections. Thermal processing techniques can significantly enhance the crystal quality of the silicon and thereby lead to higher efficiencies of the final solar cells. Further advancement into geometric considerations of building devices can exploit the dimensionality of nanomaterials. Creating large, parallel nanowire arrays enables long absorption lengths along the length of the wire while still maintaining short minority carrier diffusion lengths along the radial direction. Adding nanoparticles between the nanowires will allow for conduction through the device. Because of the natural geometry of these arrays, a textured surface will naturally form which allows for even more light to be trapped. A further advantage of this geometry is that these types of devices require about 100 times less material than conventional wafer-based devices.

Multijunction Cells

High-efficiency multijunction cells were originally developed for special applications such as satellites and space exploration, but are also now used effectively with terrestrial solar concentrators. Multijunction cells consist of multiple thin films, each essentially a solar cell in its own right, grown on top of each other, typically using metalorganic vapour phase epitaxy. A triple-junction cell, for example, may consist of the semiconductors: GaAs, Ge, and $GaInP_2$. Each type of semiconductor will have a characteristic band gap energy which, loosely speaking, causes it to absorb light most efficiently at a certain colour, or more precisely, to absorb electromagnetic radiation over a portion of the spectrum. Combinations of semiconductors are carefully chosen to efficiently absorb most of the solar spectrum, thus generating electricity from as much of the solar energy as possible.

GaAs based multijunction devices are the most efficient solar cells to date. In October 15, 2012, triple junction metamorphic cell reached a record high of 44%.

Tandem solar cells based on monolithic, series connected, gallium indium phosphide (GaInP), gallium arsenide GaAs, and germanium Ge p–n junctions, are seeing demand rapidly rise. Between December 2006 and December 2007, the cost of 4N gallium metal rose from about \$350 per kg to \$680 per kg. Additionally, germanium metal prices have risen substantially to \$1000–1200 per kg this year. Those materials include gallium (4N, 6N and 7N Ga), arsenic (4N, 6N and 7N) and germanium, pyrolitic boron nitride (pBN) crucibles for growing crystals, and boron oxide, these products are critical to the entire substrate manufacturing industry.

Triple-junction GaAs solar cells were also being used as the power source of the Dutch four-time World Solar Challenge winners Nuna in 2003, 2005 and 2007, and also by the Dutch solar cars Solutra (2005), Twente One (2007) and 21Revolution (2009).

Emerging Research Topics and Pre-industrial Technology in Solar Cells

Light-Absorbing Dyes (DSSC)

Dye-sensitized solar cells (DSSCs) are made of low-cost materials and do not need elaborate equipment to manufacture, so they can be made in a DIY fashion, possibly allowing players to produce more of this type of solar cell than others. In bulk it should be significantly less expensive than older solid-state cell designs. DSSC's can be engineered into flexible sheets, and although its conversion efficiency is less than the best thin film cells, its price/performance ratio should be high enough to allow them to compete with fossil fuel electrical generation.

Typically a ruthenium metalorganic dye (Ru-centred) is used as a monolayer of light-absorbing material. The dye-sensitized solar cell depends on a mesoporous layer of nanoparticulate titanium dioxide to greatly amplify the surface area (200–300 m^2/g TiO_2, as compared to approximately 10 m^2/g of flat single crystal). The photogenerated electrons from the *light absorbing dye* are passed on to the *n-type* TiO_2, and the holes are absorbed by an electrolyte on the other side of the dye. The circuit is completed by a redox couple in the electrolyte, which can be liquid or solid. This type of cell allows a more flexible use of materials, and is typically manufactured by screen printing or use of ultrasonic nozzles, with the potential for lower processing costs than those used for *bulk* solar cells. However, the dyes in these cells also suffer from degradation under heat and UV light, and the cell casing is difficult to seal due to the solvents used in assembly. In spite of the above, this is a popular emerging technology with some commercial impact forecast within this decade. The first commercial shipment of DSSC solar modules occurred in July 2009 from G24i Innovations.

Quantum Dot Solar Cells (QDSCs)

Quantum dot solar cells are an emerging field in solar cell research that uses quantum dots as the absorbing photovoltaic material, as opposed to better-known bulk materials such as silicon, copper indium gallium selenide (CIGS) or CdTe. Quantum dots have bandgaps that are tunable across a wide range of energy levels by changing the quantum dot size. This is in contrast to bulk materials, where the bandgap is fixed by the choice of material composition. This property makes quantum dots attractive for multi-junction solar cells, where a variety of different bandgap materials are used to improve efficiency by harvesting select portions of the solar spectrum.

The potential benefits of the quantum dot approach has led to widespread research in the field. Early examples used costly molecular beam epitaxy processes, but alternative inexpensive fabrication methods have been developed. These attempts rely on quantum dot synthesis using wet chemistry (colloidal quantum dots – CQDs) and subsequent solution processing of quantum dots. Using this approach, concentrated nanoparticle solutions are made stable by long hydrocarbon ligands which keep the nanocrystals suspended in solution. When a quantum dot solid is ready to be made, these solutions are cast down and the long stabilizing ligands are replaced with short-chain crosslinkers. Recent efforts have used chemical engineering at the nanocrystal surface to better passivated the nanocrystals and reduce detrimental trap states which would curtail device performance by means of carrier recombination. These cells result in an efficiency of 7.0% which is among the highest observed for QDSCs and, although quite low compared to that of commercial bulk silicon cells (about 17%), it has a potential for improvement beyond silicon cells by means of hot carrier extraction and efficient multiple exciton generation.

In a conventional solar cell, light is absorbed by a semiconductor producing an electron-hole (e-h) pair; the electron hole pair may be bound and is referred to as an exciton. This pair is separated by an internal electric field (present in p-n junctions or Schottcky diodes) and the resulting flow of electrons and holes creates electric current. The internal electric field is created by doping one part of semiconductor interface with atoms which act as electron donors (n-type doping) and another with electron acceptors (p-type doping) that results in a p-n junction. Generation of an e-h pair requires that the photons of light have energy exceeding the bandgap of the material. Effectively, photons with lower energies than the bandgap do not get absorbed (depends on whether the semiconductor is direct or indirect, and the exact absorption coefficients), while higher energy photons are relatively inefficient as an energetic e-h pair can quickly (within about 10^{-13} s) thermalize to the band edges, thus reducing the energy of the extracted carriers. The former limitation results in a lower current scenario, while the thermalization reduces the maximum achievable cell voltage. As a result, all semiconductor solar cells suffer a trade-off between voltage and current (which can be in part alleviated by using multiple junction implementations). The detailed balance calculation shows that this efficiency can not exceed 31% if one uses a single material for a solar cell.

Numerical analysis shows that the 31% efficiency is achieved when the solar cell material has a bandgap of 1.3-1.4 eV, corresponding to light in the near infrared. This band gap is close to that of silicon (1.1 eV), one of the many reasons this material dominates solar cell production. However, silicon's maximum efficiency is maximized at about 29%. It is possible to greatly improve on a single-junction cell by stacking cells with different bandgaps on top of each other – termed the "tandem cell" or "multi-junction" approach. The same basic analysis shows that a two layer cell should have one layer tuned to 1.64 eV and the other at 0.94 eV, with a theoretical performance of 44%. A three-layer cell should be tuned to 1.83, 1.16 and 0.71 eV, with an efficiency of 48%. An "infinity-layer" cell would have a theoretical efficiency of 86%, with other thermodynamic loss mechanisms accounting for the rest.

Traditional (crystalline) silicon preparation methods do not lend themselves to this approach due to lack of bandgap tunability. There has been some progress using thin-films of amorphous silicon, which due to a relaxed requirement in crystal momentum preservation can achieve direct bandgaps and intermixing of carbon can tune the bandgap, but other issues have prevented these from matching the performance of traditional cells. Most tandem-cell structures are based on higher performance semiconductors, notably gallium arsenide (GaAs). Three-layer InGaAs/GaAs/InGaP cells (bandgaps 1.89/1.42/0.94 eV) hold the efficiency record of 42.3% for experimental examples.

Current Research

Early Concepts: The idea of using quantum dots as a path to high efficiency was first noted by Burnham and Duggan in 1990. At the time, the science of quantum dots, or "wells" as they were known, was in its infancy and early production examples were just becoming available.

DSSC Efforts

Another modern cell design is the dye-sensitized solar cell, or DSSC. DSSCs use a sponge-like layer of TiO_2 as the semiconductor valve as well as a mechanical support structure. During construction, the sponge is filled with an organic dye, typically ruthenium-polypyridine, which injects electrons into the titanium dioxide upon photoexcitation. This dye is relatively expensive, and ruthenium is a rare metal. Another drawback of the design is that it requires direct contact between the dye molecules suspended in the film and the rear electrode to return electrons to the dye. In most designs, this is handled by a liquid electrolyte between the two, making the design susceptible to leakage currents if the electrolyte makes direct contact with the titanium dioxide. Finally, in order to efficiently harvest all the solar light, the front electrode has to be transparent. Such an electrode is usually made of indium tin oxide (ITO), which contains expensive indium metal. Recently, other conductive doped oxide materials, such as fluorite tin oxide, have been proposed to alleviate this issue.

Using quantum dots as an alternative to the molecular dyes was considered from the earliest days of DSSC research. The ability to tune the bandgap means the designer can select a wider variety materials for other portions of the cell. The collaborating groups from the University of Toronto and École Polytechnique Fédérale de Lausanne developed a new design based on a rear electrode directly in contact with a film of quantum dots, eliminating the electrolyte and forming a depleted heterojunction. To date these cells have reached 7.0% certified efficiency, better than the best solid-state DSSC devices, but still below those based on liquid electrolytes.

Multi-Junction Efforts

During this period, other teams were working with nanocrystals of other semiconductors, notably cadmium telluride (CdTe). A colloidal suspension of these crystals is spin-cast onto a suitable substrate, often a thin glass slide, potted in a conductive polymer. These did not use quantum dots, but had a number of features in common with them. In particular, the method of casting a thin layer of crystals would work just as well as with quantum dots, and the use of a thin film conductor would both be applicable with few changes. In low scale production quantum dots are more expensive to form than mass-producing of nanocrystals, but the crystals are based on rare metals that are already subject to major price swings, whereas a wide variety of materials can be used to make suitable dots. As well, Cadmium and Tellurium are highly toxic materials.

Experiments using a variety of CQDs materials with spin-casting techniques started at the Sargent Group in the mid-2000s. In one noteworthy experiment, the group used lead sulfide as an infrared-sensitive electron donor to produce the highest-efficiency IR solar cells ever built. The true advantage of this technique, however, is that it yields the prospect of combining the quantum dots inherent tunability with a simple manufacturing process to allow the construction of "tandem" cells of greatly reduced cost. The cells use a rear layer of gold as an electrode, but recent experiments have shown that nickel works just as well. This would greatly reduce the cost of the system in large-scale production.

Hot-Carrier Capture

Another way to improve efficiency is to capture the extra energy in the electron when emitted from a single-bandgap material. In traditional materials like silicon, the distance from the emission site to the electrode where they are harvested is too far to allow this to occur; the electron will undergo many interactions with the crystal materials and lattice, giving up this extra energy as heat. There was great hope in the 1980s that thin films of silicon or other materials would avoid this, and capture some of this extra energy. These films are amorphous, and in practice the defects that are inherent to these materials overwhelmed this advantage. Modern thin-film cells are generally less efficient than traditional silicon.

In the case of quantum dots, or other nanostructured donors, it is possible to cast cells as uniform films that avoid the problems with defects. These would still be subject to other issues inherent to quantum dots in general, notably resistivity issues and heat retention. It appears there has been little active development along these lines.

Multiple Exciton Generation

In 2004, Los Alamos National Laboratory reported for the first time spectroscopic evidence that several excitons could be efficiently generated upon absorption of a single, energetic photon in a quantum dot. This opens up the possibility of a different approach to the same problems that tandem cells attempt to solve, capturing more of the energy in highly energetic visible photons in sunlight. In this approach, known as "carrier multiplication" (CM) or "multiple exciton generation" (MEG), the quantum dot is tuned to release multiple electron-hole pairs at a lower energy instead of just one high-energy electron-hole pair.

This increases the cell efficiency through increased photocurrent. The dots in LANL's example were made from lead selenide.

In 2010, a team at the University of Wyoming demonstrated similar performance using cells based on the DCCS design. In their examples, PbS quantum dots demonstrated two-electron ejection when the incoming photons had about three times the base bandgap energy.

NREL maintains an active research effort developing silicon-based quantum dots. In order to become cost effective, any new solar cell design will have to compete with the existing silicon industry. Silicon is plentiful and inexpensive in bulk form, it is only the processing that makes it expensive. If quantum dots with suitable properties can be made from silicon, they can compete on a cost basis. In 2007, NREL demonstrated that MEG occurs in silicon quantum dots as well as in the PbS dots.

Other Issues

Although research is still at a pre-commercialization stage, in the future quantum dot based photovoltaics may offer advantages such as mechanical flexibility (as in quantum dot-polymer composite photovoltaics) as well as low cost, clean power generation and an efficiency of 65%.

Organic/Polymer Solar Cells

Organic solar cells are a relatively novel technology, yet may hold the promise of a substantial price reduction. These cells can be processed from liquid solution, hence the possibility of a simple roll-to-roll printing process, potentially leading to inexpensive, large scale production. In addition, these cells could be beneficial for some applications where mechanical flexibility and disposability are important. Current cell efficiencies are, however, very low, and practical devices are essentially non-existent.

Organic solar cells and polymer solar cells are built from thin films (typically 100 nm) of organic semiconductors including polymers, such as polyphenylene vinylene and small-molecule compounds like copper phthalocyanine (a blue or green organic pigment) and carbon fullerenes and fullerene derivatives such as PCBM. Energy conversion efficiencies achieved to date using conductive polymers are very low compared to inorganic materials. However, recent improvements have led to a NREL (National Renewable Energy Laboratory) certified efficiency of 8.3% for the Konarka Power Plastic, and organic tandem cells in 2012 reached 11.1%.

The active region of an organic device consists of two materials, one which acts as an electron donor and the other as an acceptor. When a photon is converted into an electron hole pair, typically in the donor material, distinct from most other solar cell types, the charges tend to remain bound in the form of an exciton, and are separated when the exciton diffuses to the donor-acceptor interface. The short exciton diffusion lengths of most polymer systems tend to limit the efficiency of such devices. Nanostructured interfaces, sometimes in the form of bulk heterojunctions, can improve performance.

In 2011, researchers at the Massachusetts Institute of Technology and Michigan State University developed the first highly efficient transparent solar cells that had a power efficiency close to 2% with a transparency to the human eye greater than 65%, achieved by selectively absorbing the ultraviolet and near-infrared parts of the spectrum with small-molecule compounds. Researchers at UCLA more recently developed an analogous polymer solar cell, following the same approach, that is 70% transparent and has a 4% power conversion efficiency. The efficiency limits of both opaque and transparent organic solar cells were recently outlined. These lightweight, flexible cells can be produced in bulk at a low cost, and could be used to create power generating windows.

In 2013, researchers announced polymer cells with some 3% efficiency. They used block copolymers, self-assembling organic materials that arrange themselves into distinct layers. The research focused on P3HT-b-PFTBT that separates into bands some 16 nanometres wide.

Manufacture

Because solar cells are semiconductor devices, they share some of the same processing and manufacturing techniques as other semiconductor devices such as computer and memory chips. However, the stringent requirements for cleanliness and quality control of semiconductor fabrication are more relaxed for solar cells. Most large-scale commercial solar cell factories today make screen printed poly-crystalline or single crystalline silicon solar cells.

Poly-crystalline silicon wafers are made by wire-sawing block-cast silicon ingots into very thin (180 to 350 micrometre) slices or wafers. The wafers are usually lightly p-type doped. To make a solar cell from the wafer, a surface diffusion of n-type dopants is performed on the front side of the wafer. This forms a p–n junction a few hundred nanometres below the surface.

Anti-reflection coatings, to increase the amount of light coupled into the solar cell, are typically next applied. Silicon nitride has gradually replaced titanium dioxide as the anti-reflection coating, because of its excellent surface passivation qualities. It prevents carrier recombination at the surface of the solar cell. It is typically applied in a layer several hundred nanometres thick using plasma-enhanced chemical vapor deposition (PECVD). Some solar cells have textured front surfaces that, like anti-reflection coatings, serve to increase the amount of light coupled into the cell. Such surfaces can usually only be formed on single-crystal silicon, though in recent years methods of forming them on multicrystalline silicon have been developed. The wafer then has a full area metal contact made on the back surface, and a grid-like metal contact made up of fine "fingers" and larger "bus bars" are screen-printed onto the front surface using a silver paste. The rear contact is also formed by screen-printing a metal paste, typically aluminium. Usually this contact covers the entire rear side of the cell, though in some cell designs it is printed in a grid pattern. The paste is then fired at several hundred degrees Celsius to form metal electrodes in ohmic contact with the silicon. Some companies use an additional electro-plating step to increase the cell efficiency. After the metal contacts are made, the solar cells are interconnected by flat wires or metal ribbons, and assembled into modules or "solar panels". Solar panels have a sheet of tempered glass on the front, and a polymer encapsulation on the back.

Manufacturers and Certification

National Renewable Energy Laboratory tests and validates solar technologies. There are three reliable certifications of solar equipment: UL and

IEEE (both U.S. standards) and IEC. There are a huge array of solar cell companies worldwide.

Solar cells are manufactured primarily in Japan, Germany, mainland China, Taiwan, Malaysia and the United States, whereas Europe, China, the U.S., and Japan have dominated (94% or more as of 2013) in installed systems for the last decade. Numerous other nations, however, have or are acquiring significant solar cell production capacity. While technologies are constantly evolving toward higher efficiencies, the most effective cells for low cost electrical production are not necessarily those with the highest efficiency, but those with a balance between low-cost production and efficiency high enough to minimize area-related balance of systems cost. Those companies with large scale manufacturing technology for coating inexpensive substrates may, in fact, ultimately be the lowest cost net electricity producers, even with cell efficiencies that are lower than those of single-crystal technologies.

Global PV cell/module production increased by 10% in 2012 despite a 9% decline in solar energy investments according to the annual 'PV Status Report" released by the European Commission's Joint Research Centre." Since 2009 cell production has quadrupled.

China

Due mainly to heavy government investment, China has played by far the most important role in changing the face of solar cell manufacturing in recent years. Chinese companies produced solar cells/modules with a capacity of ~23 GW in 2013 (60% of global production).

United States

Solar cell production in the U.S. has suffered due to the global financial crisis, but is now on the rise, partly due to the precipitously falling price of quality silicon.

Third Generation Photovoltaic Cell

Third generation photovoltaic cells are solar cells that are potentially able to overcome the Shockley–Queisser limit of 31-41% power efficiency for single bandgap solar cells. This includes a range of alternatives to the so-called "first generation solar cells" (which are solar cells made of semiconducting p-n junctions) and "second generation solar cells" (based on reducing the cost of first generation cells by employing thin film technologies). Common third-generation systems include multi-layer ("tandem") cells made of amorphous silicon or gallium arsenide, while more theoretical developments include frequency conversion, hot-carrier effects and other multiple-carrier ejection.

Solar cells can be thought of as visible light counterparts to radio receivers. A receiver consists of three basic parts; an antenna that converts the radio waves (light) into wave-like motions of electrons in the antenna material, an electronic valve that traps the electrons as they pop off the end of the antenna, and a tuner that amplifies electrons of a selected frequency. It is possible to build a solar cell identical to a radio, a system known as an optical rectenna, but to date these have not been practical.

Instead, the vast majority of the solar electric market is made up of silicon-based devices. In silicon cells, the silicon acts as both the antenna (or electron donor, technically) as well as the electronic valve. Silicon is almost ideal as a solar cell material; it is widely available, relatively inexpensive, and has a bandgap that is ideal for solar collection. On the downside it is energetically expensive to produce silicon in bulk, and great efforts have been made to reduce or eliminate the silicon in a cell. Moreover it is mechanically fragile, which typically requires a sheet of strong glass to be used as mechanical support and protection from the elements. The glass alone is a significant portion of the cost of a typical solar module.

According to the Shockley–Queisser limit, the majority of a cell's theoretical efficiency is due to the difference in energy between the bandgap and solar photon. Any photon with more energy than the bandgap can cause photoexcitation, but in this case any energy above and beyond the bandgap energy is lost. Consider the solar spectrum; only a small portion of the light reaching the ground is blue, but those photons have three times the energy of red light. Silicon's bandgap is 1.1 eV, about that of red light, so in this case the extra energy contained in blue light is lost in a silicon cell. If the bandgap is tuned higher, say to blue, that energy is now captured, but only at the cost of rejecting all the lower energy photons. It is possible to greatly improve on a single-junction cell by stacking extremely thin cells with different bandgaps on top of each other - the "tandem cell" or "multi-junction" approach. Traditional silicon preparation methods do not lend themselves to this approach. There has been some progress using thin-films of amorphous silicon, notably Uni-Solar's products, but other issues have prevented these from matching the performance of traditional cells. Most tandem-cell structures are based on higher performance semiconductors, notably gallium arsenide (GaAs). Three-layer GaAs cells hold the production record of 41.6% for experimental examples.

Numerical analysis shows that the "perfect" single-layer solar cell should have a bandgap of 1.13 eV, almost exactly that of silicon. Such a cell can have a maximum theoretical power conversion efficiency of 33.7% - the solar power below red (in the infrared) is lost, and the extra energy of the higher colours is also lost. For a two layer cell, one layer should be tuned to 1.64 eV and the other at 0.94 eV, with a theoretical performance of 44%. A three-layer cell should be tuned to 1.83, 1.16 and 0.71 eV, with an efficiency of 48%. A theoretical "infinity-layer" cell would have a theoretical efficiency of 64%.

Technologies

The third generation is somewhat ambiguous in the technologies that it encompasses, though generally it tends to include, among others, non-semiconductor technologies (including polymer cells and biomimetics), quantum dot, tandem/multi-junction cells, intermediate band solar cell, hot-carrier cells, photon upconversion and downconversion technologies, and solar thermal technologies, such as thermophotonics, which is one technology identified by Green as being third generation.

It also includes:

- Silicon nanostructures
- Modifying incident spectrum (concentration), to reach 300-500 suns and efficiencies of 32% (already attained in Sol3g cells) to +50%.
- Use of excess thermal generation (caused by UV light) to enhance voltages or carrier collection.
- Use of infrared spectrum to produce electricity at night.

Expected Market Shift

There has been a lot of hype circling around the possibilities of advanced solar technology in recent years. Major companies and investors such as Google, have invested hundreds of millions of dollars towards this new generation of solar power. They are counting on the likely possibility that the new technologies could compete with not only traditional solar cells, but more importantly with fossil fuels and nuclear energy (to reach and surpass grid parity). This would revolutionize our energy market; as said, in order for this to happen, third generation solar cells will need to be more efficient and cheaper. There are also further issues with durability (for solar cells to be able to reach their full lifetime in adverse outdoor conditions), safety (some solar manufacturing techniques involve dangerous chemicals), and energy storage (solar energy is only available at limited times, and almost unavailable all winter in non-tropical climates).

Future Possibilities

The new materials that solar energy can be harnessed with is one of the most exciting elements of the new technology. The flexible and lightweight physical characteristics of the different types of third generation solar cells makes many new applications possible.

There is the possibility that solar cells could be integrated into clothing which would allow us to have personal wireless power without batteries.

Another plausible application could be a type of automobile paint that is blended with polymer solar cells. This could help maintain the lightweight form of a solar car while still providing ample energy to power it.

Types of Third Generation Solar Cells

While the new solar technologies that have been discovered centre around nanotechnology, there are several different material methods currently used.

Amorphous Silicon

Amorphous silicon (a-Si) is the non-crystalline allotropic form of silicon. It can be deposited in thin films at low temperatures onto a variety of substrates. It offers some unique capabilities for a variety of electronics. Silicon is a fourfold coordinated atom that is normally tetrahedrally bonded to four neighbouring silicon atoms. In crystalline silicon (c-Si) this tetrahedral structure continues over a large range, thus forming a well-ordered crystal lattice.

In amorphous silicon this long range order is not present. Rather, the atoms form a continuous random network. Moreover, not all the atoms within amorphous silicon are fourfold coordinated. Due to the disordered nature of the material some atoms have a dangling bond. Physically, these dangling bonds represent defects in the continuous random network and may cause anomalous electrical behaviour.

Likewise, the material can be passivated by hydrogen, which bonds to the dangling bonds and can reduce the dangling bond density by several orders of magnitude. Hydrogenated amorphous silicon (a-Si:H) has a sufficiently low amount of defects to be used within devices such as solar photovoltaic cells, particularly in the protocrystalline growth regime. However, hydrogenation is unfortunately associated with light-induced degradation of the material, termed the Staebler–Wronski effect.

Amorphous Silicon and Carbon

Amorphous alloys of silicon and carbon (amorphous silicon carbide, also hydrogenated, a-$Si_{1-x}C_x$:H) are an interesting variant. Introduction of carbon atoms adds extra degrees of freedom for control of the properties of the material. The film could also be made transparent to visible light.

Increasing concentrations of carbon in the alloy widen the electronic gap between conduction and valence bands (also called "optical gap" and bandgap). This can potentially increase the light efficiency of solar cells made with amorphous silicon carbide layers. On the other hand, the electronic properties as a semiconductor (mainly electron mobility), are adversely affected by the increasing content of carbon in the alloy, due to the increased disorder in the atomic network.

Several studies are found in the scientific literature, mainly investigating the effects of deposition parameters on electronic quality, but practical applications of amorphous silicon carbide in commercial devices are still lacking.

Applications

While a-Si suffers from lower electronic performance compared to c-Si, it is much more flexible in its applications. For example, a-Si layers can be made thinner than c-Si, which may produce savings on silicon material cost.

One further advantage is that a-Si can be deposited at very low temperatures, e.g., as low as 75 degrees Celsius. This allows for deposition on not only glass, but plastic as well, making it a candidate for a roll-to-roll processing technique. Once deposited, a-Si can be doped in a fashion similar to c-Si, to form p-type or n-type layers and ultimately to form electronic devices.

Another advantage is that a-Si can be deposited over large areas by PECVD. The design of the PECVD system has great impact on the production cost of such panel, therefore most equipment suppliers put their focus on the design of PECVD for higher throughput, that leads to lower manufacturing cost particularly when the silane is recycled.

Amorphous silicon has become the material of choice for the active layer in thin-film transistors (TFTs), which are most widely used in large-area electronics applications, mainly for liquid-crystal displays (LCDs).

Solar Cells

a-Si has been used as a photovoltaic solar cell material for devices which require very little power, such as pocket calculators, because their lower performance compared to traditional c-Si solar cells is more than offset by their simplified and lower cost of deposition onto a substrate.

More recently, improvements in a-Si construction techniques have made them more attractive for large-area solar cell use as well. Here their lower inherent efficiency is made up, at least partially, by their thinness – higher efficiencies can be reached by stacking several thin-film cells on top of each other, each one tuned to work well at a specific frequency of light. This approach is not applicable to c-Si cells, which are thick as a result of their construction technique and are therefore largely opaque, blocking light from reaching other layers in a stack.

The main advantage of a-Si in large scale production is not efficiency, but cost. a-Si cells use approximately 1% of the silicon needed for typical c-Si cells, and the cost of the silicon is by far the largest factor in cell cost. However, the higher costs of manufacture due to the multi-layer construction have, to date, made a-Si unattractive except in roles where their thinness or flexibility are an advantage.

Typically, amorphous silicon thin-film cells use a p-i-n structure. Typical panel structure includes front side glass, TCO, thin film silicon, back contact, polyvinyl butyral (PVB) and back side glass. UNI-SOLAR, a division of Energy

Conversion Devices produces a version of flexible backings, used in roll-on roofing products.

Photovoltaic Thermal Hybrid Solar Collectors

Photovoltaic thermal hybrid solar collectors, sometimes known as hybrid PV/T systems or PVT, are systems that convert solar radiation into thermal and electrical energy. These systems combine a photovoltaic cell, which converts electromagnetic radiation (photons) into electricity, with a solar thermal collector, which captures the remaining energy and removes waste heat from the PV module. The capture of both electricity and heat allow these devices to have higher energy and thus be more overall energy efficient than solar photovoltaic (PV) or solar thermal alone. A significant amount of research has gone into developing PVT technology since the 1970s.

Photovoltaic cells suffer from a drop in efficiency with the rise in temperature due to increased resistance. Such systems can be engineered to carry heat away from the PV cells thereby cooling the cells and thus improving their efficiency by lowering resistance. Although this is an effective method, it causes the thermal component to under-perform compared to a solar thermal collector. Recent research showed that photovoltaic materials with low temperature coefficients such as amorphous silicon (a-Si:H) PV allow the PVT to be operated at high temperatures, creating a more symbiotic PVT system. This advantage can be tuned by controlling the dispatch strategy of thermal annealing cycles.

System Types

A number of PV/T collectors in different categories are commercially available and can be divided into the following categories:

- PV/T liquid collector
- PV/T air collector
- PV/Ta Liquid and air collector
- PV/T concentrator (CPVT)

PV/T Liquid Collector

The basic water-cooled design uses conductive-metal piping or plates attached to the back of a PV module. The fluid flow arrangement through the cooling element will determine which systems the panels are most suited to.

In a standard fluid based system, a working fluid, typically water, glycol or mineral oil is then piped through these pipes or plate chillers. The heat from the PV cells are conducted through the metal and absorbed by the working fluid (presuming that the working fluid is cooler than the operating temperature of the cells). In closed-loop systems this heat is either exhausted (to cool it), or transferred at a heat exchanger, where it flows to its application. In open-loop systems, this heat is used, or exhausted before the fluid returns to the PV cells. It is also possible to disperse nanoparticles in the liquid to create a liquid filter for PV/T applications. The basic advantage of this type of split configuration is that the thermal collector and the photovoltaic collector can operate at different temperatures.

Microcrystalline and Micromorphous Silicon

Microcrystalline silicon (also called nanocrystalline silicon) is amorphous silicon, but also contains small crystals. It absorbs a broader spectrum of light and is flexible.

Micromorphous silicon module technology combines two different types of silicon, amorphous and microcrystalline silicon, in a top and a bottom photovoltaic cell. Sharp produces cells using this system in order to more efficiently capture blue light, increasing the efficiency of the cells during the time where there is no direct sunlight falling on them. Protocrystalline silicon is often used to optimize the open circuit voltage of a-Si photovoltaics.

Micromorph

The artificial word micromorph is a combination of the words MICROcrystalline and aMORPHous. It is used for a type of thin-film silicon multijunction solar cell.

In this Multijunction photovoltaic cell the top cell, consisting of a thin amorphous silicon layer, absorbing the blue light and a thicker microcrystalline silicon bottom cell absorbing the red and near-infrared light of the solar spectrum.

Since the bandgaps of amorphous Silicon (1.7eV) and microcrystalline Silicon (1.1eV) are well suited for tandem solar cells, the Shockley-Queisser limit of this cell allows conversion efficiencies of over 30%. In reality this limit can not be reached and typical stable efficiencies are about 9% (world record 11.7%). That is well over the stable efficiencies of single junction thin film silicon solar cells which are around 6%. One reason of the low costs of silicon thin film solar cells is its very low thickness (2 μm) compared to silicon wafer (200 μm). In the red and infrared wavelength range 2 μm of silicon are not enough to absorb all light and therefore Light trapping is needed.

The advantage of the micromorph approach is that it keeps the thickness of the amorphous top cell low. This reduces the effect of degradation induced by light (Staebler-Wronski effect). In multijunction cells top and bottom cell have to produce the same current. But the top cell is limited by the Staebler-Wronski effect and therefore Light trapping and an intermediate reflector are needed to keep the thickness of the top cell low while increasing its current.

The Intermediate Reflector

The intermediate reflector is a layer (IRL) of Zinc oxide (ZnO intermediate reflector: ZIR) or Silicon oxide (SiOx intermediate reflector: SOIR) between the top and the bottom cell. Due to its lower refractive index of around 2 comparing to the surrounding silicon (4) light is reflected back into the top cell. This increases the top cell current from around 10mA/cm^2 to 12mA/cm^2, but reduces the bottom cell current by an equal amount.

The Word Micromorph

This artificial word was first mentioned in a scientific publication of the University of Neuchâtel research group of Prof. Arvind Shah by the author J. Meier in the year 1995 [1], but is based on long pioneering research of several authors and of several years. For a detailed list of publications.

Years later, other European, Japanese and American research groups started research activities in the field of improving the conversion efficiency of thin-film silicon solar cells by utilising the stacked solar cell concept, naming the micromorph device 'hybrid' solar cell or naming the microcrystalline silicon absorber 'nanocrystalline' or even 'policrystalline' silicon. The word 'micromorph' has been lately claimed to belong to a equipment manufacturer of silicon coating tools, but in a patent judgement, this claim has not been accepted by European patent offices [2].

Large-Scale Production

Xunlight Corporation, which has received over $40 million of institutional investments, has completed the installation of its first 25 MW wide-web, roll-to-roll photovoltaic manufacturing equipment for the production of thin-film silicon PV modules. Anwell Technologies has also completed the installation of its first 40 MW a-Si thin film solar panel manufacturing facility in Henan with its in-house designed multi-substrate-multi-chamber PECVD equipment.

Nanocrystal Solar Cell

Nanocrystal solar cells are solar cells based on a substrate with a coating of nanocrystals. The nanocrystals are typically based on silicon, CdTe or CIGS and the substrates are generally silicon or various organic conductors. Quantum dot solar cells are a variant of this approach, but take advantage of quantum mechanical effects to extract further performance. Dye-sensitized solar cells are another related approach, but in this case the nano-structuring is part of the substrate.

Previous fabrication methods relied on expensive molecular beam epitaxy processes, but colloidal synthesis allows for cheaper manufacture. A thin film of nanocrystals is obtained by a process known as "spin-coating". This involves placing an amount of the quantum dot solution onto a flat substrate, which is then rotated very quickly. The solution spreads out uniformly, and the substrate is spun until the required thickness is achieved.

Quantum dot based photovoltaic cells based around dye-sensitized colloidal TiO_2 films were investigated in 1991 and were found to exhibit promising efficiency of converting incident light energy to electrical energy, and to be incredibly encouraging due to the low cost of materials used. A single-nanocrystal (channel) architecture in which an array of single particles between the electrodes, each separated by ~1 exciton diffusion length, was proposed to improve the device efficiency and research on this type of solar cell is being conducted by groups at Stanford, Berkeley and the University of Tokyo.

Although research is still in its infancy, in the future nanocrystal photovoltaics may offer advantages such as flexibility (quantum dot-polymer composite photovoltaics) lower costs, clean power generation and an efficiency of 65%, compared to around 27% for first-generation photovoltaics.

It is argued that many measurements of the efficiency of the nanocrystal solar cell are incorrect and that nanocrystal solar cells are not suitable for large scale manufacturing.

Recent research has experimented with lead selenide (PbSe) semiconductor, as well as with cadmium telluride (CdTe), which has already been well established in the production of "classic" solar cells. Other materials are being researched as well.

7

Optoelectronics

Optoelectronics is the study and application of electronic devices that source, detect and control light, usually considered a sub-field of photonics. In this context, *light* often includes invisible forms of radiation such as gamma rays, X-rays, ultraviolet and infrared, in addition to visible light.

Optoelectronic devices are electrical-to-optical or optical-to-electrical transducers, or instruments that use such devices in their operation. *Electro-optics* is often erroneously used as a synonym, but is in fact a wider branch of physics that deals with all interactions between light and electric fields, whether or not they form part of an electronic device.

Optoelectronics is based on the quantum mechanical effects of light on electronic materials, especially semiconductors, sometimes in the presence of electric fields.

- Photoelectric or photovoltaic effect, used in:
  - photodiodes (including solar cells)
  - phototransistors
  - photomultipliers
  - integrated optical circuit (IOC) elements
- Photoconductivity, used in:
  - photoresistors
  - photoconductive camera tubes
  - charge-coupled imaging devices

- Stimulated emission, used in:
  - o injection laser diodes
  - o quantum cascade lasers
- Lossev effect, or radiative recombination, used in:
  - o light-emitting diodes or LED
  - o OLEDs
- Photoemissivity, used in
  - o photoemissive camera tube

Important applications of optoelectronics include:

- Optocoupler
- Optical fibre communications

Polycrystalline Silicon

Polycrystalline silicon, also called polysilicon, is a material consisting of small silicon crystals. It differs from single-crystal silicon, used for electronics and solar cells, and from amorphous silicon, used for thin film devices and solar cells.

Single Crystalline vs Polycrystalline Silicon

In single crystal silicon, the crystalline framework is homogenous, which can be recognised by an even external colouring. In single crystal silicon, also called monocrystal, the crystal lattice of the entire sample is continuous and unbroken with no grain boundaries. Large single crystals are exceedingly rare in nature and can also be difficult to produce in the laboratory. In contrast, in an amorphous structure the order in atomic positions is limited to short range.

Polycrystalline and paracrystalline phases are composed of a number of smaller crystals or *crystallites*. Polycrystalline silicon (or semicrystalline silicon, polysilicon, poly-Si, or simply "poly") is a material consisting of multiple small silicon crystals. Polycrystalline cells can be recognised by a visible grain, a "metal flake effect". Semiconductor grade (also solar grade) polycrystalline silicon is converted to "single crystal" silicon – meaning that the randomly associated crystallites of silicon in "polycrystalline silicon" are converted to a large "single" crystal. Single crystal silicon is used to manufacture most Si-based microelectronic devices. Polycrystalline silicon can be as much as 99.9999% pure. Ultra-pure poly is used in the semiconductor industry, starting from poly rods that are two to three metres in length. In microelectronic industry (semiconductor industry), poly is used both at the macro-scale and micro-scale (component) level. Single crystals are grown using the Czochralski process, float-zone and Bridgeman techniques.

Polycrystalline Silicon Components

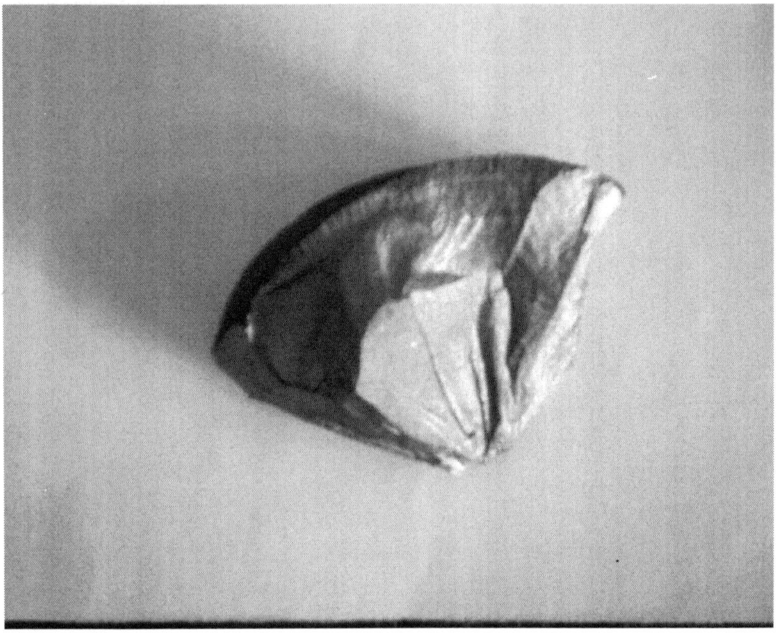

Figure: *Polycrystalline silicon (used to produce silicon monocrystals by Czochralski process)*

At the component level, polysilicon has long been used as the conducting gate material in MOSFET and CMOS processing technologies. For these technologies it is deposited using low-pressure chemical-vapour deposition (LPCVD) reactors at high temperatures and is usually heavily doped n-type or p-type.

More recently, intrinsic and doped polysilicon is being used in large-area electronics as the active and/or doped layers in thin-film transistors. Although it can be deposited by LPCVD, plasma-enhanced chemical vapour deposition (PECVD), or solid-phase crystallization (SPC) of amorphous silicon in certain processing regimes, these processes still require relatively high temperatures of at least 300 °C. These temperatures make deposition of polysilicon possible for glass substrates but not for plastic substrates. The deposition of polycrystalline silicon on plastic substrates is motivated by the desire to be able to manufacture digital displays on flexible screens. Therefore, a relatively new technique called laser crystallization has been devised to crystallize a precursor amorphous silicon (a-Si) material on a plastic substrate without melting or damaging the plastic. Short, high-intensity ultraviolet laser pulses are used to heat the deposited a-Si material to above the melting point of silicon, without melting the entire substrate. The molten silicon will then crystallize as it cools.

By precisely controlling the temperature gradients, researchers have been able to grow very large grains, of up to hundreds of micrometres in size in the extreme case, although grain sizes of 10 nanometres to 1 micrometre are also common. In order to create devices on polysilicon over large-areas however, a crystal grain size smaller than the device feature size is needed for homogeneity of the devices. Another method to produce poly-Si at low temperatures is metal-induced crystallization where an amorphous-Si thin film can be crystallized at temperatures as low as 150 °C if annealed while in contact of another metal film such as aluminium, gold, or silver.

Figure: *A polycrystalline silicon rod made by the Siemens process*

Polysilicon has many applications in VLSI manufacturing. One of its primary uses is as gate electrode material for MOS devices. A polysilicon gate's electrical conductivity may be increased by depositing a metal (such as tungsten) or a metal silicide (such as tungsten silicide) over the gate. Polysilicon may also be employed as a resistor, a conductor, or as an ohmic contact for shallow junctions, with the desired electrical conductivity attained by doping the polysilicon material.

One major difference between polysilicon and a-Si is that the mobility of the charge carriers of the polysilicon can be orders of magnitude larger and the material also shows greater stability under electric field and light-induced stress. This allows more complex, high-speed circuity to be created on the glass substrate along with the a-Si devices, which are still needed for their low-leakage characteristics. When polysilicon and a-Si devices are used in the same process this is called hybrid processing. A complete polysilicon active layer process is also used in some cases where a small pixel size is required, such as in projection displays.

Solar Panel

A solar panel is a set of solar photovoltaic modules electrically connected and mounted on a supporting structure. A photovoltaic module is a packaged, connected assembly of solar cells. The solar panel can be used as a component of a larger photovoltaic system to generate and supply electricity in commercial and residential applications.

Figure: *An installation of 24 solar modules in rural Mongolia*

Figure: *A solar photovoltaic module is composed of individual PV cells. This crystalline-silicon module has an aluminium frame and glass on the front.*

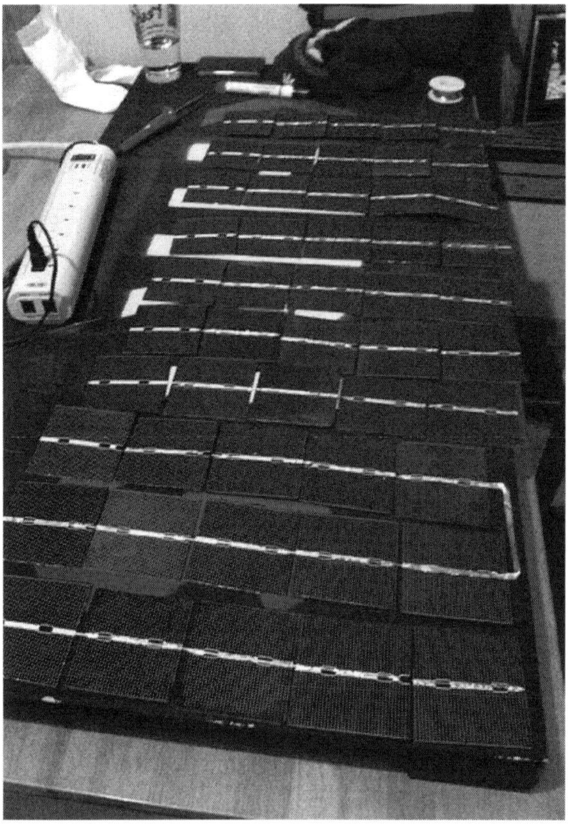

Figure: *A half-built homemade solar module, made from individual cells soldered together*

Each module is rated by its DC output power under standard test conditions (STC), and typically ranges from 100 to 320 watts. The efficiency of a module determines the area of a module given the same rated output - an 8% efficient 230 watt module will have twice the area of a 16% efficient 230 watt module. A single solar module can produce only a limited amount of power; most installations contain multiple modules. A photovoltaic system typically includes a panel or an array of solar modules, an inverter, and sometimes a battery and/or solar tracker and interconnection wiring.

Solar modules use light energy (photons) from the sun to generate electricity through the photovoltaic effect. The majority of modules use wafer-based crystalline silicon cells or thin-film cells based on cadmium telluride or silicon. The structural (load carrying) member of a module can either be the top layer or the back layer. Cells must also be protected from mechanical damage and moisture. Most solar modules are rigid, but semi-flexible ones are available, based on thin-film cells. These early solar modules were first used in space in 1958.

Electrical connections are made in series to achieve a desired output voltage and/or in parallel to provide a desired current capability. The conducting wires that take the current off the modules may contain silver, copper or other non-magnetic conductive transition metals. The cells must be connected electrically to one another and to the rest of the system. Externally, popular terrestrial usage photovoltaic modules use MC3 (older) or MC4 connectors to facilitate easy weatherproof connections to the rest of the system.

Bypass diodes may be incorporated or used externally, in case of partial module shading, to maximize the output of module sections still illuminated.

Some recent solar module designs include concentrators in which light is focused by lenses or mirrors onto an array of smaller cells. This enables the use of cells with a high cost per unit area (such as gallium arsenide) in a cost-effective way.

Efficiencies

Depending on construction, photovoltaic modules can produce electricity from a range of frequencies of light, but usually cannot cover the entire solar range (specifically, ultraviolet, infrared and low or diffused light). Hence much of the incident sunlight energy is wasted by solar modules, and they can give far higher efficiencies if illuminated with monochromatic light. Therefore, another design concept is to split the light into different wavelength ranges and direct the beams onto different cells tuned to those ranges. This has been projected to be capable of raising efficiency by 50%.

Currently the best achieved sunlight conversion rate (solar module efficiency) is around 21.5% in new commercial products typically lower than the efficiencies of their cells in isolation. The most efficient mass-produced solar modules have energy density values of up to 175 W/m^2 (16.22 W/ft^2). A research by Imperial College, London has shown that the efficiency of a solar panel can be improved by studding the light-receiving semiconductor surface with aluminum nanocylinders similar to the ridges on Lego blocks. The scattered light then travels along a longer path in the semiconductor which meant that more photons could be absorbed and converted into current. Although these nanocylinders were used previously in which aluminum was preceded by gold and silver, the light scattering occurred in the near infrared region and visible light was absorbed strongly. Aluminum was found to have absorbed ultraviolet part of the spectrum and the visible and near infrared parts of the spectrum were found to be scattered by the aluminum surface. This, the research argued, could bring down the cost significantly and improve the efficiency as aluminum is more abundant and less costly than gold and silver. The research also noted that the increase in current makes thinner film solar panels technically feasible without "compromising power conversion efficiencies, thus reducing material consumption".

Micro-inverted solar panels are wired in parallel which produces more output than normal panels which are wired in series with the output of the series determined by the lowest performing panel (this is known as the "Christmas light effect"). Micro-inverters work independently so each panel contributes its maximum possible output given the available sunlight.

Rigid Thin-Film Modules

In rigid thin film modules, the cell and the module are manufactured in the same production line.

The cell is created on a glass substrate or superstrate, and the electrical connections are created *in situ*, a so-called "monolithic integration". The substrate or superstrate is laminated with an encapsulant to a front or back sheet, usually another sheet of glass.

The main cell technologies in this category are CdTe, or a-Si, or a-Si+uc-Si tandem, or CIGS (or variant). Amorphous silicon has a sunlight conversion rate of 6-12%.

Flexible Thin-Film Modules

Flexible thin film cells and modules are created on the same production line by depositing the photoactive layer and other necessary layers on a flexible substrate.

If the substrate is an insulator (e.g. polyester or polyimide film) then monolithic integration can be used.

If it is a conductor then another technique for electrical connection must be used.

The cells are assembled into modules by laminating them to a transparent colourless fluoropolymer on the front side (typically ETFE or FEP) and a polymer suitable for bonding to the final substrate on the other side. The only commercially available (in MW quantities) flexible module uses amorphous silicon triple junction (from Unisolar).

So-called inverted metamorphic (IMM) multijunction solar cells made on compound-semiconductor technology are just becoming commercialized in July 2008. The University of Michigan's solar car that won the North American Solar Challenge in July 2008 used IMM thin-film flexible solar cells.

The requirements for residential and commercial are different in that the residential needs are simple and can be packaged so that as solar cell technology progresses, the other base line equipment such as the battery, inverter and voltage sensing transfer switch still need to be compacted and unitized for residential use. Commercial use, depending on the size of the service will be limited in the photovoltaic cell arena, and more complex parabolic reflectors and solar concentrators are becoming the dominant technology.

Flexible thin-film panels are optimal for portable applications as they are much more resistant to breakage than regular crystalline cells, but can be broken by bending them into a sharp angle. They are also much lighter per square foot than standard rigid solar panels.

The global flexible and thin-film photovoltaic (PV) market, despite caution in the overall PV industry, is expected to experience a CAGR of over 35% to 2019, surpassing 32 GW according to a major new study by IntertechPira.

Smart Solar Modules

Several companies have begun embedding electronics into PV modules. This enables performing maximum power point tracking (MPPT) for each module individually, and the measurement of performance data for monitoring and fault detection at module level. Some of these solutions make use of power optimizers, a DC-to-DC converter technology developed to maximize the power harvest from solar photovoltaic systems. As of about 2010, such electronics can also compensate for shading effects, wherein a shadow falling across a section of a module causes the electrical output of one or more strings of cells in the module to fall to zero, but not having the output of the entire module fall to zero.

Module Performance and Aging

Module performance is generally rated under standard test conditions (STC): irradiance of 1,000 W/m^2, solar spectrum of AM 1.5 and module temperature at 25°C.

Electrical characteristics include nominal power (P_{MAX}, measured in W), open circuit voltage (V_{OC}), short circuit current (I_{SC}, measured in amperes), maximum power voltage (V_{MPP}), maximum power current (I_{MPP}), peak power, W_p, and module efficiency (%).

Nominal voltage refers to the voltage of the battery that the module is best suited to charge; this is a leftover term from the days when solar modules were only used to charge batteries. The actual voltage output of the module changes as lighting, temperature and load conditions change, so there is never one specific voltage at which the module operates. Nominal voltage allows users, at a glance, to make sure the module is compatible with a given system.

Open circuit voltage or V_{OC} is the maximum voltage that the module can produce when not connected to an electrical circuit or system. V_{OC} can be measured with a metre directly on an illuminated module's terminals or on its disconnected cable.

The peak power rating, W_p, is the maximum output under standard test conditions (not the maximum possible output). Typical modules, which could measure approximately 1x2 metres or 2x4 feet, will be rated from as low as

75 watts to as high as 350 watts, depending on their efficiency. At the time of testing, the test modules are binned according to their test results, and a typical manufacturer might rate their modules in 5 watt increments, and either rate them at +/- 3%, +/-5%, +3/-0% or +5/-0%.

Solar modules must withstand rain, hail, and cycles of heat and cold for many years. Many crystalline silicon module manufacturers offer a warranty that guarantees electrical production for 10 years at 90% of rated power output and 25 years at 80%.

Recycling

Most parts of a solar module can be recycled including up to 97% of certain semiconductor materials or the glass as well as large amounts of ferrous and non-ferrous metals. Some private companies and non-profit organisations are currently engaged in take-back and recycling operations for end-of-life modules.

Recycling possibilities depend on the kind of technology used in the modules:

- Silicon based modules: aluminium frames and junction boxes are dismantled manually at the beginning of the process. The module is then crushed in a mill and the different fractions are separated - glass, plastics and metals. It is possible to recover more than 80% of the incoming weight. This process can be performed by flat glass recyclers since morphology and composition of a PV module is similar to those flat glasses used in the building and automotive industry. The recovered glass for example is readily accepted by the glass foam and glass insulation industry.
- Non-silicon based modules: they require specific recycling technologies such as the use of chemical baths in order to separate the different semiconductor materials. For cadmium telluride modules, the recycling process begins by crushing the module and subsequently separating the different fractions. This recycling process is designed to recover up to 90% of the glass and 95% of the semiconductor materials contained. Some commercial-scale recycling facilities have been created in recent years by private companies.

Since 2010, there is an annual European conference bringing together manufacturers, recyclers and researchers to look at the future of PV module recycling.

Solar Panel and Applications

Polycrystalline silicon is also a key component of solar panel construction. Growth of the photovoltaic solar industry was limited by the supply of the polysilicon material. For the first time, in 2006, over half of the world's supply of polysilicon was being used for production of renewable electricity solar power panels. Only twelve factories were known to produce solar-grade polysilicon in 2008, however by 2013 the number now stands at over 100 manufacturers. Monocrystalline silicon is higher priced and more efficient than polycrystalline.

Deposition Methods

Polysilicon deposition, or the process of depositing a layer of polycrystalline silicon on a semiconductor wafer, is achieved by pyrolyzing silane (SiH_4) at 580 to 650 °C. This pyrolysis process releases hydrogen.

Polysilicon layers can be deposited using 100% silane at a pressure of 25–130 Pa (0.19–0.98 Torr) or with 20–30% silane (diluted in nitrogen) at the same total pressure. Both of these processes can deposit polysilicon on 10–200 wafers per run, at a rate of 10–20 nm/min and with thickness uniformities of ±5%. Critical process variables for polysilicon deposition include temperature, pressure, silane concentration, and dopant concentration. Wafer spacing and load size have been shown to have only minor effects on the deposition process. The rate of polysilicon deposition increases rapidly with temperature, since it follows Arrhenius behaviour, that is deposition rate = $A \cdot \exp(-qE_a/kT)$ where q is electron charge and k is the Boltzmann constant. The activation energy (E_a) for polysilicon deposition is about 1.7 eV. Based on this equation, the rate of polysilicon deposition increases as the deposition temperature increases. There will be a minimum temperature, however, wherein the rate of deposition becomes faster than the rate at which unreacted silane arrives at the surface. Beyond this temperature, the deposition rate can no longer increase with temperature, since it is now being hampered by lack of silane from which the polysilicon will be generated. Such a reaction is then said to be 'mass-transport-limited.' When a polysilicon deposition process becomes mass-transport-limited, the reaction rate becomes dependent primarily on reactant concentration, reactor geometry, and gas flow.

When the rate at which polysilicon deposition occurs is slower than the rate at which unreacted silane arrives, then it is said to be surface-reaction-limited. A deposition process that is surface-reaction-limited is primarily dependent on reactant concentration and reaction temperature. Deposition processes must be surface-reaction-limited because they result in excellent thickness uniformity and step coverage. A plot of the logarithm of the deposition rate against the reciprocal of the absolute temperature in the surface-reaction-limited region results in a straight line whose slope is equal to $-qE_a/k$.

At reduced pressure levels for VLSI manufacturing, polysilicon deposition rate below 575 °C is too slow to be practical. Above 650 °C, poor deposition uniformity and excessive roughness will be encountered due to unwanted gas-phase reactions and silane depletion. Pressure can be varied inside a low-pressure reactor either by changing the pumping speed or changing the inlet gas flow into the reactor. If the inlet gas is composed of both silane and nitrogen, the inlet gas flow, and hence the reactor pressure, may be varied either by changing the nitrogen flow at constant silane flow, or changing both the nitrogen and silane flow to change the total gas flow while keeping the gas ratio constant. Recent investigations have shown that e-beam evaporation, followed by SPC (if needed) can be a cost effective and faster alternative for producing solar grade poly-Si thin films. Modules produced by such method are shown to have an photovoltaic efficiency of ~6%.

Polysilicon doping, if needed, is also done during the deposition process, usually by adding phosphine, arsine, or diborane. Adding phosphine or arsine results in slower deposition, while adding diborane increases the deposition rate. The deposition thickness uniformity usually degrades when dopants are added during deposition.

Upgraded Metallurgical-Grade Silicon

Upgraded metallurgical-grade (UMG) silicon (also known as UMG-Si) solar cell is being produced as a low cost alternative to polysilicon created by the Siemens process. UMG greatly reduces impurities in a variety of ways that require less equipment and energy than the Siemens process. UMG is about 99% pure which is three or more orders of magnitude less pure and about 10 times less expensive than polysilicon ($1.70 to $3.20 per kg from 2005 to 2008 compared to $40 to $400 per kg for polysilicon). It has the potential to provide nearly-as-good solar cell efficiency at 1/5 the capital expenditure, half the energy requirements, and less than $15/kg.

In 2008 several companies were touting the potential of UMG in 2010, but the credit crisis greatly lowered the cost of polysilicon and several UMG producers put plans on hold. The Siemens process will remain the dominant form of production for years to come due to more efficiently implementing the Siemens process. GT Solar claims a new Siemens process can produce at $27/kg and may reach $20/kg in 5 years. GCL-Poly expects production costs to be $20/kg by end of 2011. Elkem Solar estimates their UMG costs to be $25/kg, with a capacity of 6,000 tonnes by the end of 2010.

Calisolar expects UMG technology to produce at $12/kg in 5 years with boron at 0.3 ppm and phosphorus at 0.6 ppm. At $50/kg and 7.5 g/W, module manufacturers spend $0.37/W for the polysilicon. For comparison, if a CdTe manufacturer pays spot price for tellurium ($420/kg in April 2010) and has a 3 μm thickness, their cost would be 10 times less, $0.037/Watt. At 0.1 g/W and $31/ozt for silver, polysilicon solar producers spend $0.10/W on silver.

Q-Cells, Canadian Solar, and Calisolar have used Timminco UMG. Timminco is able to produce UMG-Si with 0.5 ppm boron for $21/kg but were sued by shareholders because they had expected $10/kg. RSI and Dow Corning have also been in litigation over UMG-Si technology.

Capacity

The polysilicon manufacturing market is growing very fast. According to Digitimes, in July 2011, the total polysilicon production in 2010 was 209,000 tons. First-tier suppliers account for 64% of the market while China-based polysilicon firms have 30% of market share.

The total production is likely to increase 37.4% to 281,000 tons by end of 2011. For 2012, EETimes Asia predicts 328,000 tons production with only 196,000 tons of demand, with spot prices expected to fall 56%. While good for renewable energy prospects, the subsequent drop in price could be brutal for manufacturers. As of late 2012, SolarIndustryMag reports a capacity of 385,000 tons will be reached by yearend 2012.

But as established producers (mentioned below) expand their capacities, additional newcomers – many from Asia – are moving into the market. Even long-time players in the field have recently had difficulties expanding plant production. It is yet unclear which companies will be able to produce at costs low enough to be profitable after the steep drop in spot-prices of the last months.

Price

Prices of polysilicon are often divided into two categories, contract and spot prices, and higher purity commands higher prices. While in booming installation times, price rally occurs in polysilicon. Not only spot prices surpass contract prices in the market; but it is also hard to acquire enough polysilicon. Buyers will accept down payment and long term agreements to acquire a large enough volume of polysilicon.

On the contrary, spot prices will be below contract prices once the solar PV installation is in a down trend. In late 2010, booming installation brought up the spot prices of polysilicon. In the first half of 2011, prices of polysilicon kept strong owing to the FIT policies of Italy. The solar PV price survey and market research firm, PVinsights, reported that the prices of polysilicon might be dragged down by lack of installation in the second half of 2011. As recently as 2008 prices were over $400/kg spiking from levels around $200/kg, while seen falling to $15/kg in 2013.

Dumping

The Chinese government accused United States and South Korean manufacturers of predatory pricing or "dumping". As a consequence, it imposed import tariffs of as much as 57 percent on polysilicon shipped from these two countries in order to stop the product from being sold below cost.

Photovoltaic Systems

A photovoltaic system (informally, PV system) is an arrangement of components designed to supply usable electric power for a variety of purposes, using the Sun (or, less commonly, other light sources) as the power source.

PV systems may be built in various configurations:

- Off-grid without battery (array-direct)
- Off-grid with battery storage for DC-only appliances
- Off-grid with battery storage for AC and DC appliances
- Grid-tie without battery
- Grid-tie with battery storage

A photovoltaic array (also called a solar array) consists of multiple photovoltaic modules, casually referred to as solar panels, to convert solar radiation (sunlight) into usable direct current (DC) electricity. A photovoltaic system for residential, commercial, or industrial energy supply normally contains an array of photovoltaic (PV) modules, one or more DC to alternating current (AC) power converters (also known as inverters), a racking system that supports the solar modules, electrical wiring and interconnections, and mounting for other components. Optionally, a photovoltaic system may include any or all of the following: renewable energy credit revenue-grade metre, maximum power point tracker (MPPT), battery system and charger, GPS solar tracker, energy management software, solar concentrators, solar irradiance sensors, anemometre, or task-specific accessories designed to meet specialised requirements for a system owner. The number of modules in the system determines the total DC watts capable of being generated by the solar array; however, the inverter ultimately governs the amount of AC watts that can be distributed for consumption. For example: A PV system comprising 11 kilowatts DC (kWDC) worth of PV modules, paired with one 10-kilowatt AC (kWAC) inverter, will be limited by the maximum output of the inverter: 10 kW AC.

A small PV system is capable of providing enough AC electricity to power a single home, or even an isolated device in the form of AC or DC electric. For example, military and civilian Earth observation satellites, street lights, construction and traffic signs, electric cars, solar-powered tents, and electric aircraft may contain integrated photovoltaic systems to provide a primary or auxiliary power source in the form of AC or DC power, depending on the design and power demands.

Large grid-connected photovoltaic power systems are capable of providing an energy supply for multiple consumers. The electricity generated can be stored, used directly (island/standalone plant), fed into a large electricity grid powered by central generation plants (grid-connected or grid-tied plant), or combined with one, or many, domestic electricity generators to feed into a small electrical grid (hybrid plant). PV systems are generally designed in order to ensure the highest energy yield for a given investment.

In the United States, the Authority Having Jurisdiction (AHJ) will review designs and issue permits, before construction can lawfully begin. Electrical installation practices must comply with standards set forth within the National Electrical Code (NEC) and be inspected by the AHJ to ensure compliance with building code, electrical code, and fire safety code. Jurisdictions may require that equipment has been tested, certified, listed, and labelled by at least one of the Nationally Recognised Testing Laboratories (NRTL). Despite the complicated installation process, a recent list of solar contractors shows a majority of installation companies were founded since 2000.

Components

Silicon Boule & Solar Cell

In order to make a Monocrystalline solar cell, a silicon ingot, also known as a silicon *boule (crystal)*, must first be produced. Once a silicon ingot has been made, it is thinly sliced and semiconductors are imbedded in the disk. The silicon disk will have positive and negative leads, which serve as connection points to tie multiple cells in series. Once multiple cells are connected in series, the formation of a module begins. Other types of solar cells are available.

Amorphous Silicon Cell

Although this cell has the same elements as a monocrystalline solar cell, the main difference between each is the degradation period as well as price to produce the cell. Although technology is supposed to improve this technology, the monocrystalline cell is considered better, as there has been more advances and applications used for this type of cell.

Photovoltaic Modules

Due to the low voltage of an individual solar cell (typically ca. 0.5V), several cells are wired in series in the manufacture of a "laminate". The laminate is assembled into a protective weatherproof enclosure, thus making a photovoltaic module or solar panel. Modules may then be strung together into a photovoltaic array.

Photovoltaic Arrays

A *photovoltaic array* (or *solar array*) is a linked collection of solar panels. The power that one module can produce is seldom enough to meet requirements of a home or a business, so the modules are linked together to form an *array*. Most PV arrays use an inverter to convert the DC power produced by the modules into alternating current that can power lights, motors, and other loads. The modules in a PV array are usually first connected in series to obtain the desired voltage; the individual strings are then connected in parallel to allow the system to produce more current. Solar panels are typically measured under STC (standard test conditions) or PTC (PVUSA test conditions), in watts. Typical panel ratings range from less than 100 watts to over 400 watts. The array rating consists of a summation of the panel ratings, in watts, kilowatts, or megawatts.

Mounting Systems

Modules are assembled into arrays on some kind of mounting system, which may be classified as ground mount, roof mount or pole mount. For solar parks a large rack is mounted on the ground, and the modules mounted on the rack. For buildings, many different racks have been devised for pitched roofs. For flat roofs, racks, bins and building integrated solutions are used. Solar panel racks mounted on top of poles can be stationary or moving. Side-of-pole mounts are suitable for situations where a pole has something else mounted at its top, such as a light fixture or an antenna. Pole mounting raises what would otherwise be a ground mounted array above weed shadows and livestock, and may satisfy electrical code requirements regarding inaccessibility of exposed wiring. Pole mounted panels are open to more cooling air on their underside, which increases performance. A multiplicity of pole top racks can be formed into a parking carport or other shade structure. A rack which does not follow the sun from left to right may allow seasonal adjustment up or down.

Trackers

A solar tracker tilts a solar panel throughout the day. Depending on the type of tracking system, the panel is either aimed directly at the sun or the brightest area of a partly clouded sky. Trackers greatly enhance early morning and late afternoon performance, increasing the total amount of power produced by a system by about 20–25% for a single axis tracker and about 30% or more for a dual axis tracker, depending on latitude. Trackers are effective in regions that receive a large portion of sunlight directly. In diffuse light (i.e. under cloud or fog), tracking has little or no value. Because most concentrated photovoltaics systems are very sensitive to the sunlight's angle, tracking systems allow them to produce useful power for more than a brief period each day. Tracking systems improve performance for two main reasons. First, when a solar panel is perpendicular to the sunlight, it receives more light on its surface than if it were angled. Second, direct light is used more efficiently than angled light. Special Anti-reflective coatings can improve solar panel efficiency for direct and angled light, somewhat reducing the benefit of tracking.

Inverters

Systems designed to deliver alternating current (AC), such as grid-connected applications need an inverter to convert the direct current (DC) from the solar modules to AC. Grid connected inverters must supply AC electricity in sinusoidal form, synchronized to the grid frequency, limit feed in voltage to no higher than the grid voltage and disconnect from the grid if the grid voltage is turned off. Islanding inverters need only produce regulated voltages and frequencies in a sinusoidal waveshape as no synchronisation or co-ordination with grid supplies is required. A solar inverter may connect to a string of solar panels. In some installations a solar micro-inverter is connected at each solar panel. For safety reasons a circuit breaker is provided both on the AC and DC side to enable maintenance. AC output may be connected through an electricity metre into the public grid.

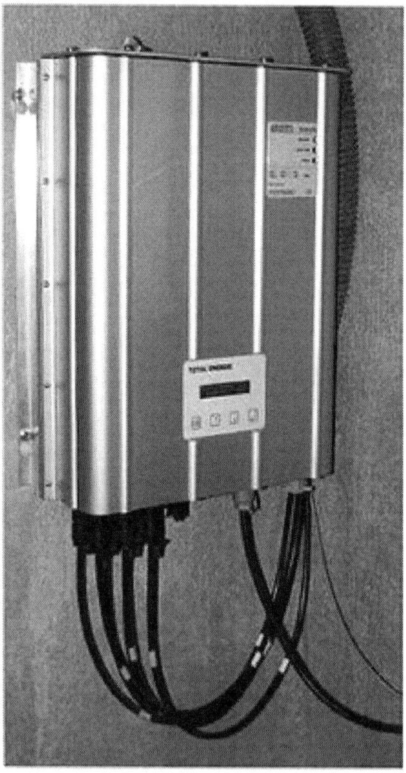

Figure: *Inverter for grid connected PV*

Maximum Power Point Tracking and Charge Control

Maximum power point tracking (MPPT) is used to maximize module output power. The power output of a module varies as a function of the voltage in a way that power generation can be optimized by varying the system voltage to find the 'maximum power point'. Some inverters incorporate maximum power point tracking.

In the case of PV systems which include a battery, a charge controller is needed to adjust the constantly varying voltage and current available from

PV panels, to correctly charge the battery. Basic charge controllers may simply turn the PV panels on and off, or may metre out pulses of energy as needed, a strategy called PWM or pulse-width modulation. More advanced charge controllers will incorporate MPPT logic into their battery charging algorithms. Charge controllers may also divert energy to some purpose other than battery charging. Rather than simply shut off the free PV energy when not needed, a user may choose to heat air or water once the battery is full.

A charge controller with MPPT capability frees the system designer from closely matching available PV voltage to battery voltage. Considerable efficiency gains can be achieved, particularly when the PV array is located at some distance from the battery. By way of example, a 150 volt PV array connected to an MPPT charge controller can be used to charge a 24 or 48 volt battery. Higher array voltage means lower array current, so the savings in wiring costs can more than pay for the controller.

Monitoring and Metring

The metring must be able to accumulate energy units in both directions or two metres must be used. Many metres accumulate bidirectionally, some systems use two metres, but a unidirectional metre (with detent) will not accumulate energy from any resultant feed into the grid.

In some countries, for installations over 30kWp a frequency and a voltage monitor with disconnection of all phases is required. This is done to pre solar power is being generated than can be accommodated by the utility, and can not either be exported or stored. Grid operators historically have needed to provide transmission lines and generation capacity. Now they need to also provide storage. This is normally hydro-storage, but other means of storage are used. Initially storage was used so that baseload generators could operate at full output. With variable renewable energy, storage is needed to allow power generation whenever it is available, and consumption whenever it is needed. The two variables a grid operator have are storing electricity for *when* it is needed, or transmitting it to *where* it is needed. If both of those fail, installations over 30kWp can automatically shut down, although in practice all inverters maintain voltage regulation and stop supplying power if the load is inadequate. Grid operators have the option of curtailing excess generation from large systems, although this is more commonly done with wind power than solar power, and results in a substantial loss of revenue. Inverters have the unique option of supplying reactive power which can be advantageous in matching load requirements.

Standalone Applications

Stand-alone photovoltaic power systems are electrical power systems energised by photovoltaic panels which are independent of the utility grid.

These types of systems may use solar panels only or may be used in conjunction with a diesel generator or a wind turbine.

Types

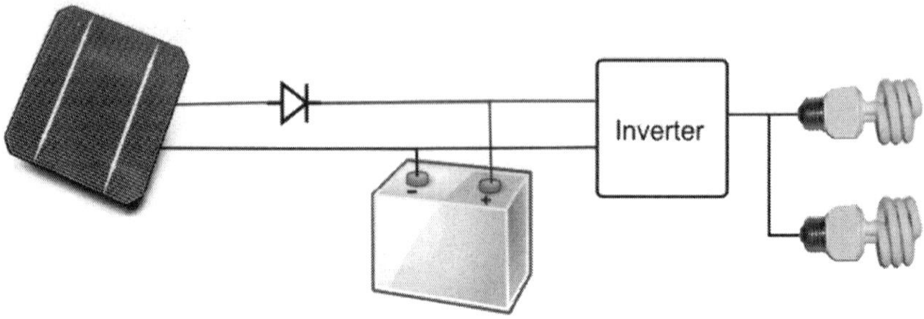

Figure: *A schematic of a direct-coupled system, consisting (from left to right) of PV module, battery, inverter, and AC load*

The two types of stand-alone photovoltaic power systems are direct-coupled system without batteries and stand alone system with batteries.

Direct-Coupled System

The basic model of a direct coupled system consists of a solar panel connected directly to a dc load. As there are no battery banks in this setup, energy is not stored and hence it is capable of powering common appliances like fans, pumps etc. only during the day. MPPTs are generally used to efficiently utilise the Sun's energy especially for electrical loads like positive-displacement water pumps. Impedance matching is also considered as a design criteria in direct-coupled systems.

Stand Alone System with Batteries

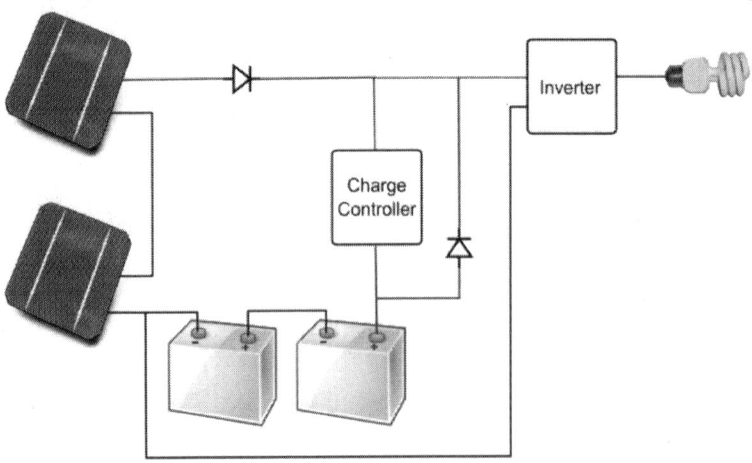

Figure: *A typical stand-alone photovoltaic power system with battery and battery charger arrangement*

In stand-alone photovoltaic power systems, the electrical energy produced by the photovoltaic panels cannot always be used directly. As the demand from the load does not always equal the solar panel capacity, battery banks are generally used. The primary functions of a storage battery in a stand-alone PV system are:

- Energy Storage Capacity and Autonomy: To store energy when there is an excees is available and to provide it when required.
- Voltage and Current Stabilization: To provide stable current and voltage by eradicating transients.
- Supply Surge Currents: to provide surge currents to loads like motors when required.

System monitoring

Monitoring photovoltaic systems can provide useful information about their operation and what should be done to improve performance, but if the data are not reported properly, the effort is wasted. To be helpful, a monitoring report must provide information on the relevant aspects of the operation in terms that are easily understood by a third party. Appropriate performance parameters need to be selected, and their values consistently updated with each new issue of the report. In some cases it may be beneficial to monitor the performance of individual components in order to refine and improve system performance, or be alerted to loss of performance in time for preventative action. For example, monitoring battery charge/discharge profiles will signal when replacement is due before downtime from system failure is experienced.

IEC Standard 61724

IEC has provided a set of monitoring standards called the "Standard for Photovoltaic system performance monitoring". It focusses on the photovoltaic system's electrical performance and it does not address hybrids or prescribe a method for ensuring that performance assessments are equitable.

Performance Assessment

Performance assessment involves:

- Data collection, which is a straightforward process of measuring parameters.
- Evaluation of that data in a manner that provides useful information.
- Dissemination of useful information to the end user.

Load Related Problems

The wide range of load related problems identified are classified into the following types:

- Wrong selection: Some loads cannot be used with stand-alone PV systems.
- House wiring: Inadequate or low quality wiring and protection devices can affect the system's response.
- Low efficiency: Low efficiency loads may increase energy consumption.
- Stand-by loads: Stand-by mode of some loads waste energy.
- Start-up: High current drawn by some loads during start-up. Current spikes during the start-up can overload the system temporarily.
- Reactive power: The circulating current can differ from the current consumed when capacitive or inductive loads are used.
- Harmonic distortion: Non-linear loads may create distortion of the inverter waveform.
- Mismatch between load and inverter size: When a higher rated inverter is used for a lower capacity load, overall efficiency is reduced.

Pumps

Solar well pumps are common and widespread. They often meet a need for water beyond the reach of power lines, taking the place of a windmill or windpump. One common application is the filling of livestock watering tanks, so that grazing cattle may drink. Another is the refilling of drinking water storage tanks on remote or self-sufficient homes.

Fans

Connecting a photovoltaic panel directly to a DC mechanical fan motor can provide air movement when it is most needed during the day. Common applications include both attic and greenhouse ventilation. Increased efficiency can be obtained by interposing a linear current booster (LCB) between the solar panel and the fan motor, to more closely coordinate varying panel output with motor energy requirements. Other controls sometimes used are timers and thermostats, so that the fan does not run when not wanted, even if the sun is shining.

Solar Vehicles

Ground, water, air or space vehicles may obtain some or all of the energy required for their operation from the sun. Surface vehicles generally require higher power levels than can be sustained by a practically sized solar array, so a battery is used to meet peak power demand, and the solar array recharges it. Space vehicles have successfully used solar photovoltaic systems for years of operation, eliminating the weight of fuel or primary batteries.

There are limits to using photovoltaic (PV) cells for vehicles:

- Power density: Power from a solar array is limited by the size of the vehicle and area that can be exposed to sunlight. While energy can be accumulated in batteries to lower peak demand on the array and provide operation in sunless conditions, the battery adds weight and cost to the vehicle. The power limit can be mitigated by use of conventional electric cars supplied by solar (or other) power, recharging from the electrical grid.

- Cost: While sunlight is free, the creation of PV cells to capture that sunlight is expensive. Costs for solar panels are steadily declining (22% cost reduction per doubling of production volume).
- Design considerations: Even though sunlight has no lifespan, PV cells do. The lifetime of a solar module is approximately 30 years. Standard photovoltaics often come with a warranty of 90% (from nominal power) after 10 years and 80% after 25 years. Mobile applications are unlikely to require lifetimes as long as building integrated PV and solar parks. Current PV panels are mostly designed for stationary installations. However, to be successful in mobile applications, PV panels need to be designed to withstand vibrations. Also, solar panels, especially those incorporating glass have significant weight. To be useful, the energy harvested by a panel must exceed the added fuel consumption caused by the added weight.

Small Scale Solar Systems

Figure: *Profile picture of a mobile solar powered generator*

With a growing DIY-community and an increasing interest in environmentally friendly "green energy", some hobbyists have endeavoured to build their own PV solar systems from kits or partly diy. Usually, the DIY-community uses inexpensive or high efficiency systems (such as those with solar tracking) to generate their own power. As a result, the DIY-systems often end up cheaper than their commercial counterparts. Often, the system is also hooked up into the regular power grid, using net metring instead of a battery for backup. These systems usually generate power amount of ~2 kW or less. Through the internet, the community is now able to obtain plans to construct the system (at least partly DIY) and there is a growing trend toward building them for domestic requirements. Small scale solar systems are now also being used both in developed countries and in developing countries, for residences and small businesses. One of the most cost effective solar applications is a solar powered pump, as it is far cheaper to purchase a solar panel than it is to run power lines.

Grid-Connected Photovoltaic Power System

Grid-connected photovoltaic power systems are power systems energised by photovoltaic panels which are connected to the utility grid. Grid-connected photovoltaic power systems consist of Photovoltaic panels, MPPT, solar inverters, power conditioning units and grid connection equipment. Unlike Stand-alone photovoltaic power systems these systems seldom have batteries. When conditions are right, the grid-connected PV system supplies the excess power, beyond consumption by the connected load, to the utility grid.

Working

Figure: *A grid-connected photovoltaic power plant in Nellis Air Force Base, Nevada*

Residential grid-connected photovoltaic power systems which have a capacity less than 10 kilowatts can meet the load of most consumers. They can feed excess power to the grid, which in this case acts as a battery for the system. The feedback is done through a metre to monitor power transferred. This is called Net metring. Photovoltaic wattage may be less than average consumption, in which case the consumer will continue to purchase grid energy, but a lesser amount than previously. If photovoltaic wattage substantially exceeds average consumption, the energy produced by the panels will be much in excess of the demand. In this case, the excess power can yield revenue by selling it to the grid. Depending on their agreement with their local grid energy company, the consumer only needs to pay the cost of electricity consumed less the value of electricity generated. This will be a negative number if more electricity is generated than consumed. Additionally, in some cases, cash incentives are paid from the grid operator to the consumer.

Connection of the photovoltaic power system can be done only through an interconnection agreement between the consumer and the utility company. The agreement details the various safety standards to be followed during the connection.

Features

Solar energy gathered by photovoltaic solar panels, intended for delivery to a power grid, must be conditioned, or processed for use, by a grid-connected inverter. This inverter sits between the solar array and the grid, draws energy from each, and may be a large stand-alone unit or may be a collection of small inverters, each physically attached to individual solar panels. The inverter must monitor grid voltage, waveform, and frequency. One reason for monitoring is if the grid is dead or strays too far out of its nominal specifications, the inverter must not pass along any solar energy. An inverter connected to a malfunctioning power line will automatically disconnect in accordance with safety rules, for example UL1741, which vary by jurisdiction. Another reason for the inverter monitoring the grid is because for normal operation the inverter must synchronize with the grid waveform, and produce a voltage slightly higher than the grid itself, in order for energy to smoothly flow outward from the solar array.

Anti-Islanding

Islanding refers to the condition in which a distributed generator continues to power a location even though power from the electric utility is no longer present. Islanding can be dangerous to utility workers, who may not realise that a circuit is still powered, even though there's no power from the electrical grid. For that reason, distributed generators must detect islanding and immediately stop producing power; this is referred to as anti-islanding.

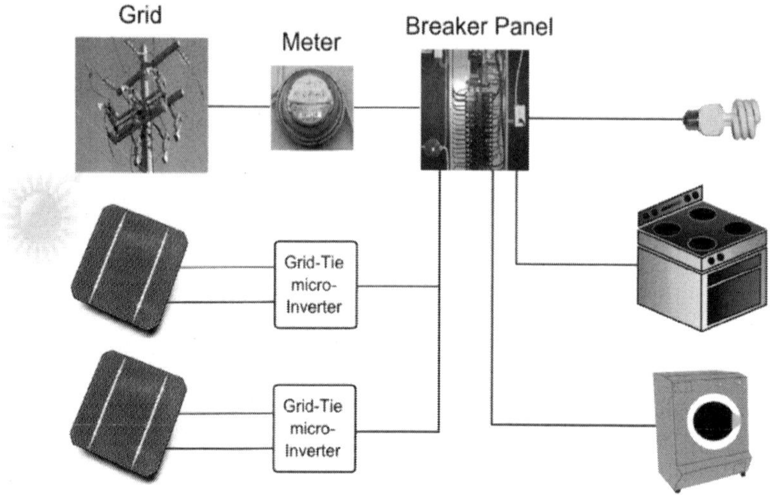

***Figure:** Diagram of a residential grid-connected PV system*

In the case of a utility blackout in a grid-connected PV system, the solar panels will continue to deliver power as long as the sun is shining. In this case, the supply line becomes an "island" with power surrounded by a "sea" of unpowered lines. For this reason, solar inverters that are designed to supply power to the grid are generally required to have automatic anti-islanding circuitry in them.

In intentional islanding, the generator disconnects from the grid, and forces the distributed generator to power the local circuit. This is often used as a power backup system for buildings that normally sell their power to the grid.

There are two types of anti-islanding control techniques:

- Passive: The voltage and/or the frequency change during the grid failure is measured and a positive feedback loop is employed to push the voltage and /or the frequency further away from its nominal value. Frequency or voltage may not change if the load matches very well with the inverter output or the load has a very high quality factor (reactive to real power ratio). So there exists some *Non Detection Zone* (NDZ).
- Active: This method employs injecting some error in frequency or voltage. When grid fails, the error accumulates and pushes the voltage and/or frequency beyond the acceptable range.

Advantages

- A grid-connected photovoltaic power system will reduce the power bill as it is possible to sell surplus electricity produced to the local electricity supplier.

- Grid-connected PV systems are comparatively easier to install as they do not require a battery system.
- Grid interconnection of photovoltaic (PV) power generation systems has the advantage of effective utilisation of generated power because there are no storage losses involved.
- A photovoltaic power system is carbon negative over its lifespan, as any energy produced over and above that to build the panel initially offsets the need for burning fossil fuels. Even though the sun doesn't always shine, any installation gives a reasonably predictable average reduction in carbon consumption.

Connection to DC Grids

DC grids are found in electric powered transport: railways trams and trolleybuses. A few pilot plants for such applications have been built, such as the tram depots in Hannover Leinhausen, using photovoltaic contributors and Geneva (Bachet de Pesay). The 150 kW_p Geneva site feeds 600V DC directly into the tram/trolleybus electricity network whereas before it provided about 15% of the electricity at its opening in 1999.

Building-Mounted and Building-Integrated Systems

In urban and suburban areas, photovoltaic arrays are commonly used on rooftops to supplement power use; often the building will have a connection to the power grid, in which case the energy produced by the PV array can be sold back to the utility in some sort of net metring agreement. Some utilities, such as Solvay Electric in Solvay, NY, use the rooftops of commercial customers and telephone poles to support their use of PV panels.

Solar trees are arrays that, as the name implies, mimic the look of trees, provide shade, and at night can function as street lights. In agricultural settings, the array may be used to directly power DC pumps, without the need for an inverter. In remote settings such as mountainous areas, islands, or other places where a power grid is unavailable, solar arrays can be used as the sole source of electricity, usually by charging a storage battery.

There is financial support available for people wishing to install PV arrays. Incentives range from federal tax credits to state tax credits and rebates to utility loans and rebates. A listing of current incentives can be found at the Database of State Incentives for Renewables and Efficiency. In the UK, households are paid a 'Feedback Fee' to buy excess electricity at a flat rate per kWh. This is up to 44.3p/kWh which can allow a home to earn double their usual annual domestic electricity bill. The current UK feed-in tariff system is due for review on 31 March 2012, after which the current scheme may no longer be available.

Ground-, Building- and Roof-Mounted PV Solar Systems Planning and Permit

While article 690 of the National Electric Code provides general guidelines for the installation of photovoltaic systems, these guidelines may be superseded by local laws and regulations. Often a permit is required necessitating plan submissions and structural calculations before work may begin. Additionally, many locales require the work to be performed under the guidance of a licensed electrician. Check with the local City/County AHJ (Authority Having Jurisdiction) to ensure compliance with any applicable laws or regulations.

Photovoltaic Power Station

A photovoltaic power station, also known as a solar park, is a large-scale photovoltaic system designed for the supply of merchant power into the electricity grid. They are differentiated from most building-mounted and other decentralised solar power applications because they supply power at the utility level, rather than to a local user or users. They are sometimes also referred to as solar farms or solar ranches, especially when sited in agricultural areas. The generic expression utility-scale solar is sometimes used to describe this type of project.

The power conversion source is via photovoltaic modules that convert light directly to electricity. This differs from the other large-scale solar generation technology, concentrated solar power. Photovoltaic power stations are typically rated in terms of the DC peak capacity of the solar arrays, in mega watt-peak (MW_p), or of the nominal maximum AC output in megawatts (MW) or mega volt-amperes (MVA). Most solar parks are developed at a scale of at least 1 MW_p. The largest operating photovoltaic power stations have capacities of hundreds of MW_p; projects up to 1 GW_p are planned. The cumulative worldwide capacity of plants of 4 MW_{AC} and over as at the end of 2013 was reported to be 21 GW.

Most of the existing large-scale photovoltaic power stations are owned and operated by independent power producers, but the involvement of community- and utility-owned projects is increasing. To date, almost all have been supported at least in part by regulatory incentives such as feed-in tariffs or tax credits, but capital costs have fallen significantly in the last decade and are expected to progressively reach grid parity, when external incentives may no longer be required.

The first 1 MW_p solar park was built by Arco Solar at Lugo near Hesperia, California at the end of 1982, followed in 1984 by a 5.2 MW_p installation in Carrizo Plain. Both have since been decommissioned, though Carrizo Plain is the site for several large plants now being constructed or planned. The next stage followed the 2004 revisions to the feed-in tariffs in Germany when a substantial volume of solar parks were constructed.

***Figure:** Serpa Solar Park built in Portugal in 2006*

Several hundred installations over 1 MW_p have been since been installed in Germany, of which more than 50 are over 10 MW_p. With its introduction of feed-in tariffs in 2008, Spain became briefly the largest market, with some 60 solar parks over 10 MW, but these incentives have since been withdrawn. The USA, China India, France, Canada, and Italy, amongst others, have also become major markets as shown on the list of photovoltaic power stations.

The largest solar parks currently operating have capacities between 200 and 300 MW_p. The largest sites under construction have capacities of hundreds of MW_p and projects at a scale of 1 GW_p are being planned. Some solar parks that are notable in the global or regional context are listed in this article.

Siting and Land Use

The land area required for solar parks varies depending on the location, and on the efficiency of the solar modules, the slope of the site and the type of mounting used. Fixed tilt solar arrays using typical modules of about 15% efficiency on horizontal sites, need about 1 hectare/MW in the tropics and this figure rises to over 2 hectares in northern Europe. Because of the longer shadow the array casts when tilted at a steeper angle, this area is typically about 10% higher for an adjustable tilt array or a single axis tracker, and 20% higher for a 2-axis tracker, though these figures will vary depending on the latitude and topography.

The best locations for solar parks in terms of land use are held to be brown field sites, or where there is no other valuable land use. Even in cultivated areas, a significant proportion of the site of a solar farm can also be devoted to other productive uses, such as grazing or biodiversity.

Co-location

In some cases several different solar power stations, with separate owners and contractors, are developed on adjacent sites. This can offer the advantage of the projects sharing the cost and risks of project infrastructure such as grid connections and planning approval. Solar farms can also be co-located with wind farms. Sometimes the title 'solar park' is used, rather than an individual solar power station.

Some examples of such solar clusters are the Charanka Solar Park, where there are 17 different generation projects; Neuhardenberg, with eleven plants, and the Golmud solar parks with total reported capacity over 500MW. An extreme example is calling all of the solar farms in the Gujarat state of India a single solar park, the Gujarat Solar Park.

Technology

Most solar parks use ground mounted (sometimes called free-field or stand-alone) arrays. They can either be fixed tilt or tracking, either single axis or dual axis. Tracking increases the output, but also the installation and maintenance cost.

The DC power output of the solar arrays is converted to AC by inverters, and connection to the grid is through a high voltage three phase step up transformer, usually to 10 kV or above.

Solar Array Arrangements

The solar arrays are the subsystems which convert incoming light into electrical energy. They comprise a multitude of solar modules, mounted on support structures and interconnected to deliver a power output to electronic power conditioning subsystems.

A minority of utility-scale solar parks are configured on buildings and so use building-mounted solar arrays. The majority are 'free field' systems using ground-mounted structures, usually of one of the following types:

Fixed Arrays

Many projects use mounting structures where the solar modules are mounted at a fixed inclination calculated to provide the optimum annual output profile. The modules are normally oriented towards the Equator, at a tilt angle slightly less than the latitude of the site. In some cases, depending on local climatic, topographical or electricity pricing regimes, different tilt angles can be used, or the arrays might be offset from the normal East-West axis to favour morning or evening output.

A variant on this design is the use of arrays, whose tilt angle can be adjusted twice or four times annually to optimise seasonal output. They also require more land area to reduce internal shading at the steeper winter tilt angle. Because the increased output is typically only a few percent, it seldom justifies the increased cost and complexity of this design.

Dual Axis Trackers

Figure: *Bellpuig Solar Park near Lerida, Spain uses pole-mounted 2-axis trackers*

To maximise the intensity of incoming direct radiation, solar panels should be orientated normal to the sun's rays. To achieve this, arrays can be designed using two-axis trackers, capable of tracking the sun in its daily orbit across the sky, and as its elevation changes throughout the year.

These arrays need to be spaced out to reduce inter-shading as the sun moves and the array orientations change, so need more land area. They also require more complex mechanisms to maintain the array surface at the required angle. The increased output can be of the order of 30% in locations with high levels of direct radiation, but the increase is lower in temperate climates or those with more significant diffuse radiation, due to overcast conditions. For this reason, dual axis trackers are most commonly used in subtropical regions, and were first deployed at utility scale at the Lugo plant.

Single Axis Trackers

A third approach achieves some of the output benefits of tracking, with a lesser penalty in terms of land area, capital and operating cost. This involves tracking the sun in one dimension - in its daily journey across the sky - but not adjusting for the seasons. The angle of the axis is normally horizontal, though some, such as the solar park at Nellis Airforce Base, which have a 20° tilt, incline the axis towards the equator in a north-south orientation - effectively a hybrid between tracking and fixed tilt.

Single axis tracking systems are aligned along axes roughly North-South. Some use linkages between rows so that the same actuator can adjust the angle of several rows at once.

Power Conversion

Solar panels produce direct current (DC) electricity, so solar parks need conversion equipment to convert this to alternating current (AC), which is the form transmitted by the electricity grid. This conversion is done by inverters. To maximise their efficiency, solar power plants also incorporate maximum power point trackers, either within the inverters or as separate units. These devices keep each solar array string close to its peak power point.

There are two primary alternatives for configuring this conversion equipment; centralised and string inverters, although in some cases individual, or micro-inverters are used. Single inverters allows optimizing the output of each panel, and multiple inverters increases the reliability by limiting the loss of output when an inverter fails.

Centralised Inverters

Figure: *Waldpolenz Solar Park is divided into blocks, each with a centralised inverter*

These units have relatively high capacity, typically of the order of 1 MW, so they condition that the output of a substantial block of solar arrays, up to perhaps 2 hectares (4.9 acres) in area. Solar parks using centralised inverters are often configured in discrete rectangular blocks, with the related inverter in one corner, or the centre of the block.

String Inverters

String inverters are substantially lower in capacity, of the order of 10 kW, and condition the output of a single array string. This is normally a whole, or part of, a row of solar arrays within the overall plant. String inverters can enhance the efficiency of solar parks, where different parts of the array are experiencing different levels of insolation, for example where arranged at different orientations, or closely packed to minimise site area.

Transformers

The system inverters typically provide power output at voltages of the order of 480 V_{AC}. Electricity grids operate at much higher voltages of the order of tens or hundreds of thousands of volts, so transformers are incorporated to deliver the required output to the grid. Due to the long lead time, the Long Island Solar Farm chose to keep a spare transformer onsite, as transformer failure would have kept the solar farm offline for a long period. Transformers typically have a life of 25 to 75 years, and normally do not require replacement during the life of a photovoltaic power station.

System Performance

The performance of a solar park is a function of the climatic conditions, the equipment used and the system configuration. The primary energy input is the global light irradiance in the plane of the solar arrays, and this in turn is a combination of the direct and the diffuse radiation. A key determinant of the output of the system is the conversion efficiency of the solar modules, which will depend in particular on the type of solar cell used.

There will be losses between the DC output of the solar modules and the AC power delivered to the grid, due to a wide range of factors such as light absorption losses, mismatch, cable voltage drop, conversion efficiencies, and other parasitic losses. A parameter called the 'performance ratio' has been developed to evaluate the total value of these losses. The performance ratio gives a measure of the output AC power delivered as a proportion of the total DC power which the solar modules should be able to deliver under the ambient climatic conditions. In modern solar parks the performance ratio should typically be in excess of 80%.

System Degradation

Early photovoltaic systems output decreased as much as 10%/year, but as of 2010 the median degradation rate was 0.5%/year, with modules made after 2000 having a significantly lower degradation rate, so that a system would lose only 12% of its output performance in 25 years. A system using modules which degrade 4%/year will lose 64% of its output during the same period. Many panel makers offer a performance guarantee, typically 90% in ten years and 80% over 25 years. The output of all panels is typically warranteed at plus or minus 3% during the first year of operation.

The Business of Developing Solar Parks

Solar power plants are developed to deliver merchant electricity into the grid as an alternative to other renewable, fossil or nuclear generating stations.

The plant owner is an electricity generator. Most solar power plants today are owned by independent power producers (IPP's), though some are held by investor- or community-owned utilities.

Some of these power producers develop their own portfolio of power plants, but most solar parks are initially designed and constructed by specialist project developers. Typically the developer will plan the project, obtain planning and connection consents, and arrange financing for the capital required. The actual construction work is normally contracted to one or more EPC (engineering, procurement and construction) contractors.

Major milestones in the development of a new photovoltaic power plant are planning consent, grid connection approval, financial close, construction, connection and commissioning. At each stage in the process, the developer will be able to update estimates of the anticipated performance and costs of the plant and the financial returns it should be able to deliver.

Planning Approval

Photovoltaic power stations occupy at least one hectare for each megawatt of rated output, so require a substantial land area; which is subject to planning approval. The chances of obtaining consent, and the related time, cost and conditions, varying from jurisdiction to jurisdiction and location to location. Many planning approvals will also apply conditions on the treatment of the site after the station has been decommissioned in the future. A professional health, safety and environment assessment is usually undertaken during the design of a PV power station in order to ensure the facility is designed and planned in accordance with all HSE regulations.

Grid Connection

The availability, locality and capacity of the connection to the grid is a major consideration in planning a new solar park, and can be a significant contributor to the cost.

Most stations are sited within a few kilometres of a suitable grid connection point. This network needs to be capable of absorbing the output of the solar park when operating at its maximum capacity. The project developer will normally have to absorb the cost of providing power lines to this point and making the connection; in addition often to any costs associated with upgrading the grid, so it can accommodate the output from the plant.

Operation and Maintenance

Before the commissioning phase, plants undergo a final installation inspection in order to ensure that the PV power station has been installed and connected in-line with all requirements. Once the solar park has been commissioned, the owner usually enters into a contract with a suitable counterparty to undertake operation and maintenance (O&M). In many cases this may be fulfilled by the original EPC contractor.

Solar parks are reliable solid-state systems requiring minimal maintenance, compared to rotating machinery for example. A major aspect of the O&M contract will be continuous monitoring of the performance of the plant and all of its primary subsystems, which is normally undertaken remotely. This enables performance to be compared with the anticipated output under the climatic conditions actually experienced. It also provides data to enable the scheduling of both rectification and preventative maintenance. A small number of large solar farms use a separate inverter or maximizer for each solar panel, which provide individual performance data that can be monitored. For other solar farms, thermal imaging is a tool that is used to identify non-performing panels for replacement.

Power Delivery

A solar park's income derives from the sales of electricity to the grid, and so its output is metred in real-time with readings of its energy output provided, typically on a half hourly basis, for balancing and settlement within the electricity market. Income is affected by the reliability of equipment within the plant and also by the availability of the grid network to which it is exporting. Some connection contracts allow the transmission system operator to constrain the output of a solar park, for example at times of low demand or high availability of other generators. Some countries make statutory provision for priority access to the grid for renewable generators, such as that under the European Renewable Energy Directive.

Economics and Finance

With the market structures and prices which have prevailed in the past photovoltaics has not been able to compete directly in the bulk electricity market. It has relied on support mechanisms such as feed-in tariffs, renewable portfolio standards, loan guarantees and tax credits. Recently costs have declined strongly towards grid parity, where it is becoming a viable source of mainstream power.

Grid Parity

Solar generating stations have become progressively cheaper in recent years, and this trend is expected to continue. Meanwhile, traditional electricity generation is becoming progressively more expensive.

These trends are expected to lead to a crossover point when the levelised cost of energy from solar parks, historically more expensive, matches the cost of traditional electricity generation. This point is commonly referred to as grid parity.

For merchant solar power stations, where the electricity is being sold into the electricity transmission network, the levelised cost of solar energy will need to match the wholesale electricity price. This point is sometimes called 'wholesale grid parity' or 'busbar parity'.

Some photovoltaic systems, such as rooftop installations, can supply power directly to an electricity user. In these cases, the installation can be competitive when the output cost matches the price at which the user pays for his electricity consumption. This situation is sometimes called 'retail grid parity', 'socket parity' or 'dynamic grid parity'.

Research carried out by UN-Energy in 2012 suggests areas of sunny countries with high electricity prices, such as Italy, Spain and Australia, and areas using diesel generators, have reached retail grid parity.

Incentive Mechanisms

Because the point of grid parity has not yet been reached in many parts of the world, solar generating stations need some form of financial incentive to compete for the supply of electricity. Many legislatures around the world have introduced such incentives to support the deployment of solar power stations.

Feed-in Tariffs

Feed in tariffs are designated prices which must be paid by utility companies for each kilowatt hour of renewable electricity produced by qualifying generators and fed into the grid. These tariffs normally represent a premium on wholesale electricity prices and offer a guaranteed revenue stream to help the power producer finance the project.

Renewable Portfolio Standards and Supplier Obligations

These standards are obligations on utility companies to source a proportion of their electricity from renewable generators. In most cases, they do not prescribe which technology should be used and the utility is free to select the most appropriate renewable sources.

There are some exceptions where solar technologies are allocated a proportion of the RPS in what is sometimes referred to as a 'solar set aside'.

Loan Guarantees and Other Capital Incentives

Some countries and states adopt less targeted financial incentives, available for a wide range of infrastructure investment, such as the US Department of Energy loan guarantee scheme, which stimulated a number of investments in the solar power plant in 2010 and 2011.

Tax Credits and Other Fiscal Incentives

Another form of indirect incentive which has been used to stimulate investment in solar power plant was tax credits available to investors. In some cases the credits were linked to the energy produced by the installations, such as the Production Tax Credits. In other cases the credits were related to the capital investment such as the Investment Tax Credits.

International, National and Regional Programmes

In addition to free market commercial incentives, some countries and regions have specific programs to support the deployment of solar energy installations.

The European Union's Renewables Directive sets targets for increasing levels of deployment of renewable energy in all member states. Each has been required to develop a National Renewable Energy Action Plan showing how these targets would be met, and many of these have specific support measures for solar energy deployment. The directive also allows states to develop projects outside their national boundaries, and this may lead to bilateral programs such as the Helios project.

The Clean Development Mechanism of the UNFCCC is an international programme under which solar generating stations in certain qualifying countries can be supported. Additionally many other countries have specific solar energy development programmes. Some examples are India's JNNSM, the Flagship Program in Australia, and similar projects in South Africa and Israel.

Financial Performance

The financial performance of the solar power plant is a function of its income and its costs.

The electrical output of a solar park will be related to the solar radiation, the capacity of the plant and its performance ratio. The income derived from this electrical output will come primarily from the sale of the electricity, and any incentive payments such as those under Feed-in Tariffs or other support mechanisms.

Electricity prices may vary at different times of day, giving a higher price at times of high demand. This may influence the design of the plant to increase its output at such times.

The dominant costs of solar power plants are the capital cost, and therefore any associated financing and depreciation. Though operating costs are typically relatively low, especially as no fuel is required, most operators will want to ensure that adequate operation and maintenance cover is available to maximise the availability of the plant and thereby optimise the income to cost ratio.

Geography

Because of the economic background, large-scale systems are presently distributed where the support regimes have been the most consistent, or the most advantageous. Activities in the key markets are reviewed individually below.

The first places to reach grid parity were those with high traditional electricity prices and high levels of solar radiation. The worldwide distribution of solar parks is expected to change as different regions achieve grid parity.

Around the World

Total capacity of worldwide PV plants above 4 MW_{AC} was assessed by Wiki-Solar as 21 GW in c. 1,500 installations at the end of 2013. The countries which had the most capacity, in descending order, were: USA, China, Germany, Spain, India, Italy and France. Based on the projects under development, this is expected to change in the future.

China

China was reported in early 2013 to have overtaken Germany as the nation with the most utility-scale solar capacity. Much of this has been supported by the Clean Development Mechanism. The distribution of power plants around the country is quite broad, with the highest concentration in the Gobi desert and connected to the Northwest China Power Grid.

Germany

The first multi-megawatt plant in Europe was the 4.2 MW community-owned project at Hemau, commissioned in 2003. But it was the revisions to the German feed-in tariffs in 2004, which gave the strongest impetus to the establishment of utility-scale solar power plants. The first to be completed under this programme was the Leipziger Land solar park developed by Geosol. Several dozen plants were built between 2004 and 2011, several of which were at the time the largest in the world.

The EEG, the law which establishes Germany's feed-in tariffs, provides the legislative basis not just for the compensation levels, but other regulatory factors, such as priority access to the grid.

The law was amended in 2010 to restrict the use of agricultural land, since which time most solar parks have been built on so-called 'development land', such as former military sites. Partly for this reason, the geographic distribution of photovoltaic power plants in Germany is biased towards the former Eastern Germany. As of February 2012, Germany had 1.1 million photovoltaic power plants (most are small kW roof mounted).

India

India has been rising up the leading nations for the installation of utility-scale solar capacity. The Charanka Solar Park in Gujarat was opened officially in April 2012 and was at the time the largest group of solar power plants in the world.

Geographically the majority of the stations are located in Gujarat and Maharashtra. Rajasthan has unsuccessfully been attempting to attract solar development. It and Gujarat share the Thar Desert, along with Pakistan.

Italy

Italy has a very large number of photovoltaic power plants, the largest of which is the 84 MW Montalto di Castro project.

Spain

The majority of the deployment of solar power stations in Spain to date occurred during the boom market of 2007-8. The stations are well distributed around the country, with some concentration in Extremadura, Castile-La Mancha and Murcia.

United Kingdom

The introduction of Feed-in tariffs in the United Kingdom in 2010 stimulated the first wave of utility-scale projects, with c. 20 plants being completed before tariffs were reduced on 1 August 2011 following the 'Fast Track Review'. A second wave of installations was undertaken under the UK's Renewables Obligation, with the total number of plants connected by the end of March 2013 reaching 86. This is reported to have made the UK Europe's best market in the first quarter of 2013.

UK projects were originally concentrated in the South West, but have more recently spread across the South of England and into East Anglia and the Midlands.

United States

The US deployment of photovoltaic power stations is largely concentrated in South Western States. The Renewable Portfolio Standards in California and surrounding states provide a particular incentive. The volume of projects under construction in early 2013 has led to the forecast that the US will become the leading market.

System Performance

Insolation and Energy

Solar insolation is made up of direct radiation, diffuse radiation and reflected radiation (or albedo). The absorption factor of a PV cell is defined as the fraction of incident solar irradiance that is absorbed by the cell. At high noon on a cloudless day at the equator, the power of the sun is about 1 kW/m^2, on the Earth's surface, to a plane that is perpendicular to the sun's rays.

As such, PV arrays can track the sun through each day to greatly enhance energy collection. However, tracking devices add cost, and require maintenance, so it is more common for PV arrays to have fixed mounts that tilt the array and face solar noon (approximately due south in the Northern Hemisphere or due north in the Southern Hemisphere).

The tilt angle, from horizontal, can be varied for season, but if fixed, should be set to give optimal array output during the peak electrical demand portion of a typical year for a stand alone system. This optimal module tilt angle is not necessarily identical to the tilt angle for maximum annual array energy output. The optimization of the a photovoltaic system for a specific environment can be complicated as issues of solar flux, soiling, and snow losses should be taken into effect.

In addition, recent work has shown that spectral effects can play a role in optimal photovoltaic material selection. For example, the spectral albedo can play a significant role in output depending on the surface around the photovoltaic system and the type of solar cell material.

For the weather and latitudes of the United States and Europe, typical insolation ranges from 4 kWh/m^2/day in northern climes to 6.5 kWh/m^2/day in the sunniest regions. Typical solar panels have an average efficiency of 15%, with the best commercially available panels at 21%.

Thus, a photovoltaic installation in the southern latitudes of Europe or the United States may expect to produce 1 kWh/m^2/day. A typical "150 watt" solar panel is about a square metre in size. Such a panel may be expected to produce 0.75 kWh every day, on average, after taking into account the weather and the latitude, for an insolation of 5 sun hours/day.

A typical 1 kW photovoltaic installation in Australia or the southern latitudes of Europe or United States, may produce 3.5-5 kWh per day, dependent on location, orientation, tilt, insolation and other factors. In the Sahara desert, with less cloud cover and a better solar angle, one could ideally obtain closer to 8.3 kWh/m^2/day provided the nearly ever present wind would not blow sand onto the units. The area of the Sahara desert is over 9 million km^2. 90,600 km^2, or about 1%, could generate as much electricity as all of the world's power plants combined.

Tracking the Sun

Trackers and sensors to optimise the performance are often seen as optional, but tracking systems can increase viable output by up to 45%. PV arrays that approach or exceed one megawatt often use solar trackers. Accounting for clouds, and the fact that most of the world is not on the equator, and that the sun sets in the evening, the correct measure of solar power is insolation – the average number of kilowatt-hours per square metre per day. For the weather and latitudes of the United States and Europe, typical insolation ranges from 2.26 $kWh/m^2/day$ in northern climes to 5.61 $kWh/m^2/day$ in the sunniest regions.

For large systems, the energy gained by using tracking systems can outweigh the added complexity (trackers can increase efficiency by 30% or more). For very large systems, the added maintenance of tracking is a substantial detriment. Tracking is not required for flat panel and low-concentration photovoltaic systems. For high-concentration photovoltaic systems, dual axis tracking is a necessity.

Pricing trends affect the balance between adding more stationary solar panels versus having fewer panels that track. When solar panel prices drop, trackers become a less attractive option.

Shading and Dirt

Photovoltaic cell electrical output is extremely sensitive to shading. The effects of this shading are well known. When even a small portion of a cell, module, or array is shaded, while the remainder is in sunlight, the output falls dramatically due to internal 'short-circuiting' (the electrons reversing course through the shaded portion of the p-n junction).

If the current drawn from the series string of cells is no greater than the current that can be produced by the shaded cell, the current (and so power) developed by the string is limited. If enough voltage is available from the rest of the cells in a string, current will be forced through the cell by breaking down the junction in the shaded portion. This breakdown voltage in common cells is between 10 and 30 volts.

Instead of adding to the power produced by the panel, the shaded cell absorbs power, turning it into heat. Since the reverse voltage of a shaded cell is much greater than the forward voltage of an illuminated cell, one shaded cell can absorb the power of many other cells in the string, disproportionately affecting panel output.

For example, a shaded cell may drop 8 volts, instead of adding 0.5 volts, at a particular current level, thereby absorbing the power produced by 16 other cells. It is, thus important that a PV installation is not shaded by trees or other obstructions. Several methods have been developed to determine shading losses from trees to PV systems over both large regions using LiDAR, but also at an individual system level using sketchup. Most modules have bypass diodes between each cell or string of cells that minimize the effects of shading and only lose the power of the shaded portion of the array.

The main job of the bypass diode is to eliminate hot spots that form on cells that can cause further damage to the array, and cause fires. Sunlight can be absorbed by dust, snow, or other impurities at the surface of the module. This can reduce the light that strikes the cells.

In general these losses aggregated over the year are small even for locations in Canada. Maintaining a clean module surface will increase output performance over the life of the module. Google found that cleaning the flat mounted solar panels after 15 months increased their output by almost 100%, but that the 5% tilted arrays were adequately cleaned by rainwater.

Temperature

Module output and life are also degraded by increased temperature. Allowing ambient air to flow over, and if possible behind, PV modules reduces this problem.

Module Efficiency

In 2012, solar panels available for consumers can have an efficiency of up to about 17%, while commercially available panels can go as far as 27%. It has been recorded that a group from the The Fraunhofer Institute for Solar Energy Systems have created a cell that can reach 44.7% efficiency, which will make scientists' hopes of reaching the 50% efficiency threshold a lot more feasible.

Monitoring

Photovoltaic systems need to be monitored to detect breakdown and optimize their operation. Several *photovoltaic monitoring* strategies depending on the output of the installation and its nature. Monitoring can be performed on site or remotely. It can measure production only, retrieve all the data from the inverter or retrieve all of the data from the communicating equipment (probes, metres, etc.).

Monitoring tools can be dedicated to supervision only or offer additional functions. Individual inverters and battery charge controllers may include monitoring using manufacturer specific protocols and software. Energy metring of an inverter may be of limited accuracy and not suitable for revenue metring purposes.

A third-party data acquisition system can monitor multiple inverters, using the inverter manufacturer's protocols, and also acquire weather-related information.

Independent smart metres may measure the total energy production of a PV array system. Separate measures such as satellite image analysis or a solar radiation metre (a pyranometre) can be used to estimate total insolation for comparison. Data collected from a monitoring system can be displayed remotely over the World Wide Web.

For example, the Open Solar Outdoors Test Field (OSOTF) is a grid-connected photovoltaic test system, which continuously monitors the output of a number of photovoltaic modules and correlates their performance to a long list of highly accurate meteorological readings. The OSOTF is organised under open source principles—All data and analysis is be made freely available to the entire photovoltaic community and the general public. The Fraunhofer Centre for Sustainable Energy Systems maintains two test systems, one in Massachusetts, and the Outdoor Solar Test Field OTF-1 in Albuquerque, New Mexico, which opened in June 2012.

A third site, OTF-2, also in Albuquerque, is under construction. Some companies offer analysis software to analyze system performance. Small residential systems may have minimal data analysis requirements other than perhaps total energy production; larger grid-connected power plants can benefit from more detailed investigations of performance.

Performance Factors

Uncertainties in revenue over time relate mostly to the evaluation of the solar resource and to the performance of the system itself. In the best of cases, uncertainties are typically 4% for year-to-year climate variability, 5% for solar resource estimation (in a horizontal plane), 3% for estimation of irradiation in the plane of the array, 3% for power rating of modules, 2% for losses due to dirt and soiling, 1.5% for losses due to snow, and 5% for other sources of error. Identifying and reacting to manageable losses is critical for revenue and O&M efficiency. Monitoring of array performance may be part of contractual agreements between the array owner, the builder, and the utility purchasing the energy produced.

Recently, a method to create "synthetic days" using readily available weather data and verification using the Open Solar Outdoors Test Field make it possible to predict photovoltaic systems performance with high degrees of accuracy. This method can be used to then determine loss mechanisms on a local scale - such as those from snow or the effects of surface coatings (e.g. hydrophobic or hydrophilic) on soiling or snow losses. Access to the Internet has allowed a further improvement in energy monitoring and communication. Dedicated systems are available from a number of vendors. For solar PV system that use microinverters (panel-level DC to AC conversion), module power data is automatically provided.

Some systems allow setting performance alerts that trigger phone/email/ text warnings when limits are reached. These solutions provide data for the system owner and the installer. Installers are able to remotely monitor multiple installations, and see at-a-glance the status of their entire installed base.

Module Life

Effective module lives are typically 25 years or more. The payback period for an investment in a PV solar installation varies greatly and is typically less useful than a calculation of return on investment. While it is typically calculated to be between 10 and 20 years, the payback period can be far shorter with incentives.

Hybrid Systems

A hybrid system combines PV with other forms of generation, usually a diesel generator. Biogas is also used. The other form of generation may be a type able to modulate power output as a function of demand. However more than one renewable form of energy may be used e.g. wind. The photovoltaic power generation serves to reduce the consumption of non renewable fuel. Hybrid systems are most often found on islands. Pellworm island in Germany and Kythnos island in Greece are notable examples (both are combined with wind). The Kythnos plant has reduced diesel consumption by 11.2%.

There has also been recent work showing that the PV penetration limit can be increased by deploying a distributed network of PV+CHP hybrid systems in the U.S. The temporal distribution of solar flux, electrical and heating requirements for representative U.S. single family residences were analyzed and the results clearly show that hybridizing CHP with PV can enable additional PV deployment above what is possible with a conventional centralized electric generation system. This theory was reconfirmed with numerical simulations using per second solar flux data to determine that the necessary battery backup to provide for such a hybrid system is possible with relatively small and inexpensive battery systems. In addition, large PV+CHP systems are possible for institutional buildings, which again provide back up for intermittent PV and reduce CHP runtime.

Standardization

Increasing use of photovoltaic systems and integration of photovoltaic power into existing structures and techniques of supply and distribution increases the value of general standards and definitions for photovoltaic components and systems. The standards are compiled at the International Electrotechnical Commission (IEC) and apply to efficiency, durability and safety of cells, modules, simulation programs, plug connectors and cables, mounting systems, overall efficiency of inverters etc.

Costs and Economy

Costs of production have been reduced dramatically in recent years for more widespread use through production and technological advances. For large-scale installations, prices below \$1.00 per watt are now common. A price decrease of 50% had been achieved in Europe from 2006 to 2011 and there is a potential to lower the generation cost by 50% by 2020. Crystal silicon solar cells have largely been replaced by less expensive multicrystalline silicon solar cells, and thin film silicon solar cells have also been developed recently at lower costs of production. Although they are reduced in energy conversion efficiency from single crystalline "siwafers", they are also much easier to produce at comparably lower costs.

The table below shows the total cost in US cents per kWh of electricity generated by a photovoltaic system. The row headings on the left show the total cost, per peak kilowatt (kW_p), of a photovoltaic installation. Photovoltaic system costs have been declining and in Germany, for example, were reported to have fallen to USD 2200/kW_p by the second quarter of 2012. The column headings across the top refer to the annual energy output in kWh expected from each installed kW_p. This varies by geographic region because the average insolation depends on the average cloudiness and the thickness of atmosphere traversed by the sunlight. It also depends on the path of the sun relative to the panel and the horizon. Panels are usually mounted at an angle based on latitude, and often they are adjusted seasonally to meet the changing solar declination. Solar tracking can also be utilised to access even more perpendicular sunlight, thereby raising the total energy output.

The calculated values in the table reflect the total cost in cents per kWh produced. They assume a 10% total capital cost (for instance 4% interest rate, 1% operating and maintenance cost, and depreciation of the capital outlay over 20 years). Normally, photovoltaic modules have a 25 year warranty.

During the cells life time, repairs will be necessary to keep the system running well. Repairs can be made on the actual structure to the sublethal effects and actions on cell membranes of near-UV, the interactions of different wavelengths, and the biological action of the mid-UV. Research has shown that some repairs have been based on irradiation in high efficiency metals which can be measured by admittance spectroscopy. These can all lead to costly repairs over time. Although these repairs can be expensive, thermal events such as the one in Bakersfield, California can occur, as faulty arrays made the system overheat and catch flame. This lead to expose more faulty areas and a larger fire.

Regulation

United Kingdom: In the UK, PV installations are generally considered permitted development and don't require planning permission. If the property is listed or in a designated area (National Park, Area of Outstanding Natural Beauty, Site of Special Scientific Interest or Norfolk Broads) then planning permission is required.

United States

In the US Many localities require a license to install a photovoltaic system. A grid-tied system normally requires a licensed electrician to make the connection between the system and the grid-connected wiring of the building. Installers who meet these qualifications are located in almost every state.

The State of California prohibits Homeowners' associations from restricting solar devices.

Spain

Although Spain uses about 20% of its energy via photovoltaics as cities such as Huelva and Seville boast nearly 3,000 hours of sunshine per year, Spain has issued a solar tax to account for the debt created by the investment done by the Spanish government. Those who do not connect to the grid can face up to a fine of 30 million euros ($40 million USD).

Space-Based Solar Power

Space-based solar power (SBSP) is the concept of collecting solar power in space (using an "SPS", that is, a "solar-power satellite" or a "satellite power system") for use on Earth. It has been in research since the early 1970s.

SBSP would differ from current solar collection methods in that the means used to collect energy would reside on an orbiting satellite instead of on Earth's surface. Some projected benefits of such a system are a higher collection rate and a longer collection period due to the lack of a diffusing atmosphere and nighttime in space.

Part of the solar energy is lost on its way through the atmosphere by the effects of reflection and absorption. Space-based solar power systems convert sunlight to microwaves outside the atmosphere, avoiding these losses, and the downtime (and cosine losses, for fixed flat-plate collectors) due to the Earth's rotation.

Besides the cost of implementing such a system, SBSP also introduces several new hurdles, primarily the problem of transmitting energy from orbit to Earth's surface for use. Since wires extending from Earth's surface to an orbiting satellite are neither practical nor feasible with current technology, SBSP designs generally include the use of some manner of wireless power transmission. The collecting satellite would convert solar energy into electrical energy on board, powering a microwave transmitter or laser emitter, and focus its beam toward a collector (rectenna) on Earth's surface. Radiation and micrometeoroid damage could also become concerns for SBSP.

In 1941, science fiction writer Isaac Asimov published the science fiction short story "Reason", in which a space station transmits energy collected from the Sun to various planets using microwave beams. The SBSP concept, originally known as satellite solar-power system (SSPS), was first described in November 1968. In 1973 Peter Glaser was granted U.S. patent number 3,781,647 for his method of transmitting power over long distances (e.g. from an SPS to Earth's surface) using microwaves from a very large antenna (up to one square kilometre) on the satellite to a much larger one, now known as a rectenna, on the ground.

Glaser then was a vice president at Arthur D. Little, Inc. NASA signed a contract with ADL to lead four other companies in a broader study in 1974. They found that, while the concept had several major problems – chiefly the expense of putting the required materials in orbit and the lack of experience on projects of this scale in space – it showed enough promise to merit further investigation and research.

Between 1978 and 1981, the Congress authorized the Department of Energy (DoE) and NASA to jointly investigate the concept. They organised the Satellite Power System Concept Development and Evaluation Program. The study remains the most extensive performed to date (budget $50 million). Several reports were published investigating the engineering feasibility of such an engineering project. They include:

- Resource Requirements (Critical Materials, Energy, and Land)

Figure: *Artist's concept of Solar Power Satellite in place. Shown is the assembly of a microwave transmission antenna. The solar power satellite was to be located in a geosynchronous orbit, 36,000 miles above the Earth's surface. NASA 1976*

- Financial/Management Scenarios
- Public Acceptance
- State and Local Regulations as Applied to Satellite Power System Microwave Receiving Antenna Facilities
- Student Participation
- Potential of Laser for SBSP Power Transmission
- International Agreements
- Centralization/Decentralization
- Mapping of Exclusion Areas For Rectenna Sites
- Economic and Demographic Issues Related to Deployment
- Some Questions and Answers
- Meteorological Effects on Laser Beam Propagation and Direct Solar Pumped Lasers
- Public Outreach Experiment
- Power Transmission and Reception Technical Summary and Assessment
- Space Transportation

The project was not continued with the change in administrations after the 1980 US Federal elections.

The Office of Technology Assessment concluded: Too little is currently known about the technical, economic, and environmental aspects of SPS to make a sound decision whether to proceed with its development and deployment. In addition, without further research an SPS demonstration or systems-engineering verification program would be a high-risk venture.

In 1997 NASA conducted its "Fresh Look" study to examine the modern state of SBSP feasibility. In assessing "What has changed" since the DOE study, NASA asserted that: US National Space Policy now calls for NASA to make significant investments in technology (not a particular vehicle) to drive the costs of ETO *[Earth to Orbit]* transportation down dramatically. This is, of course, an absolute requirement of space solar power.

Conversely, Dr. Pete Worden claimed that space-based solar is about five orders of magnitude more expensive than solar power from the Arizona desert, with a major cost being the transportation of materials to orbit. Dr. Worden referred to possible solutions as speculative, and that would not be available for decades at the earliest.

On Nov 2, 2012, China proposed space collaboration with India that mentioned SBSP, " . . . may be Space-based Solar Power initiative so that both India and China can work for long term association with proper funding along with other willing space faring nations to bring space solar power to earth."

Figure: *SERT sandwich concept.NASA*

SERT

In 1999, NASA's Space Solar Power Exploratory Research and Technology program (SERT) was initiated for the following purposes:

- Perform design studies of selected flight demonstration concepts.
- Evaluate studies of the general feasibility, design, and requirements.
- Create conceptual designs of subsystems that make use of advanced SSP technologies to benefit future space or terrestrial applications.
- Formulate a preliminary plan of action for the U.S. (working with international partners) to undertake an aggressive technology initiative.
- Construct technology development and demonstration roadmaps for critical Space Solar Power (SSP) elements.

SERT went about developing a solar power satellite (SPS) concept for a future gigawatt space power system, to provide electrical power by converting the Sun's energy and beaming it to Earth's surface, and provided a conceptual development path that would utilise current technologies. SERT proposed an inflatable photovoltaic gossamer structure with concentrator lenses or solar heat engines to convert sunlight into electricity. The program looked both at systems in sun-synchronous orbit and geosynchronous orbit.

Some of SERT's conclusions:

- The increasing global energy demand is likely to continue for many decades resulting in new power plants of all sizes being built.

- The environmental impact of those plants and their impact on world energy supplies and geopolitical relationships can be problematic.
- Renewable energy is a compelling approach, both philosophically and in engineering terms.
- Many renewable energy sources are limited in their ability to affordably provide the base load power required for global industrial development and prosperity, because of inherent land and water requirements.
- Based on their Concept Definition Study, space solar power concepts may be ready to re-enter the discussion.
- Solar power satellites should no longer be envisioned as requiring unimaginably large initial investments in fixed infrastructure before the emplacement of productive power plants can begin.
- Space solar power systems appear to possess many significant environmental advantages when compared to alternative approaches.
- The economic viability of space solar power systems depends on many factors and the successful development of various new technologies (not least of which is the availability of much lower cost access to space than has been available), however, the same can be said of many other advanced power technologies options.
- Space solar power may well emerge as a serious candidate among the options for meeting the energy demands of the 21st century. Space Solar Power Satellite Technology Development at the Glenn Research Centre—An Overview] James E. Dudenhoefer and Patrick J. George, NASA Glenn Research Centre, Cleveland, Ohio.
- Launch costs in the range of $100–$200 per kilogram of payload to low Earth orbit are needed if SPS are to be economically viable.

Advantages

The SBSP concept is attractive because space has several major advantages over the Earth's surface for the collection of solar power.

- There is no air in space, so the collecting surfaces could receive much more intense sunlight, unobstructed by the filtering effects of atmospheric gasses, cloud cover, and other weather events. Consequently, collection in orbit is approximately 144% of the maximum attainable on Earth's surface.
- A satellite could be illuminated over 99% of the time, and be in Earth's shadow on only 75 minutes per night at the spring and fall equinoxes. Orbiting satellites can be exposed to a consistently high degree of solar radiation, generally for 24 hours per day, whereas surface panels can collect for 12 hours per day at most.

- Relatively quick redirecting of power directly to areas that need it most. A collecting satellite could possibly direct power on demand to different surface locations based on geographical baseload or peak load power needs.
- Elimination of plant and wildlife interference.

Disadvantages

The SBSP concept also has a number of problems.

- The large cost of launching a satellite into space
- Inaccessibility: Maintenance of an earth-based solar panel is relatively simple, but performing maintenance on a solar panel in space incurs the extra cost of transporting a team of astronauts into space.
- The space environment is hostile; panels suffer about 8 times the degradation they would on Earth.
- Space debris is a major hazard to large objects in space, and all large structures such as SBSP systems have been mentioned as potential sources of orbital debris.
- The broadcast frequency of the microwave downlink (if used) would require isolating the SBSP systems away from other satellites. GEO space is already well used and it is considered unlikely the ITU would allow an SPS to be launched.
- The large size and corresponding cost of the receiving station on the ground.

Design

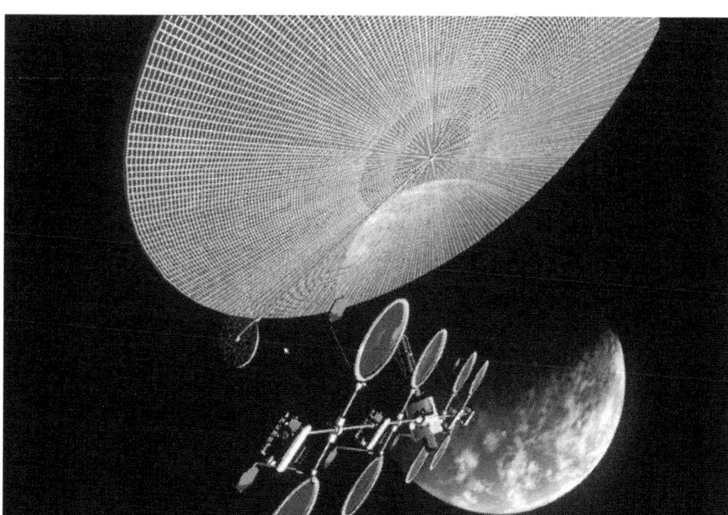

Figure: *Artist's concept of a solar disk on top of a LEO to GEO electrically powered space tug.*

Space-based solar power essentially consists of three elements:

- a means of collecting solar power in space, for example via solar concentrators, solar cells or a heat engine
- a means of transmitting power to earth, for example via microwave or laser
- a means of receiving power on earth, for example via a microwave antenna (rectenna)

The space-based portion will not need to support itself against gravity (other than relatively weak tidal stresses). It needs no protection from terrestrial wind or weather, but will have to cope with space hazards such as micrometeors and solar flares.

Two basic methods of conversion have been studied: photovoltaic (PV) and solar dynamic (SD). Photovoltaic conversion uses semiconductor cells to directly convert photons into electrical power. Solar dynamic uses mirrors to concentrate light on a boiler. The use of solar dynamic could reduce mass per watt. Most analyses of SBSP have focused on photovoltaic conversion (commonly known as "solar cells").

Wireless power transmission was proposed early on as a means to transfer energy from collection to the Earth's surface, using either microwave or laser radiation at a variety of frequencies.

Solar Concentrator

Microwave Power Transmission

William C. Brown demonstrated in 1964, during Walter Cronkite's CBS News program, a microwave-powered model helicopter that received all the power it needed for flight from a microwave beam. Between 1969 and 1975, Bill Brown was technical director of a JPL Raytheon program that beamed 30 kW of power over a distance of 1-mile (1.6 km) at 84% efficiency.

Microwave power transmission of tens of kilowatts has been well proven by existing tests at Goldstone in California (1975) and Grand Bassin on Reunion Island (1997).

More recently, microwave power transmission has been demonstrated, in conjunction with solar energy capture, between a mountain top in Maui and the main island of Hawaii (92 miles away), by a team under John C. Mankins. Technological challenges in terms of array layout, single radiation element design, and overall efficiency, as well as the associated theoretical limits are presently a subject of research, as it is demonstrated by the Special Session on "Analysis of Electromagnetic Wireless Systems for Solar Power Transmission" to be held in the 2010 IEEE Symposium on Antennas and Propagation.

In 2013, a useful overview was published, covering technologies and issues associated with microwave power transmission from space to ground. It includes an introduction to SPS, current research and future prospects.

Laser Power Beaming

Laser power beaming was envisioned by some at NASA as a stepping stone to further industrialization of space. In the 1980s, researchers at NASA worked on the potential use of lasers for space-to-space power beaming, focusing primarily on the development of a solar-powered laser. In 1989 it was suggested that power could also be usefully beamed by laser from Earth to space. In 1991 the SELENE project (SpacE Laser ENErgy) had begun, which included the study of laser power beaming for supplying power to a lunar base. The SELENE program was a two-year research effort, but the cost of taking the concept to operational status was too high, and the official project ended in 1993 before reaching a space-based demonstration.

In 1988 the use of an Earth-based laser to power an electric thruster for space propulsion was proposed by Grant Logan, with technical details worked out in 1989. He proposed using diamond solar cells operating at 600 degrees to convert ultraviolet laser light.

Orbital Location

The main advantage of locating a space power station in geostationary orbit is that the antenna geometry stays constant, and so keeping the antennas lined up is simpler. Another advantage is that nearly continuous power transmission is immediately available as soon as the first space power station is placed in orbit; other space-based power stations have much longer start-up times before they are producing nearly continuous power.

A collection of LEO (Low Earth Orbit) space power stations has been proposed as a precursor to GEO (Geostationary Orbit) space-based solar power.

Earth-Based Receiver

The Earth-based rectenna would likely consist of many short dipole antennas connected via diodes. Microwave broadcasts from the satellite would be received in the dipoles with about 85% efficiency. With a conventional microwave antenna, the reception efficiency is better, but its cost and complexity are also considerably greater. Rectennas would likely be several kilometres across.

In Space Applications

A laser SBSP could also power a base or vehicles on the surface of the Moon or Mars, saving on mass costs to land the power source. A spacecraft or another satellite could also be powered by the same means.

Dealing with Launch Costs

One problem for the SBSP concept is the cost of space launches and the amount of material that would need to be launched.

Reusable launch systems are predicted to provide lower launch costs to low Earth orbit (LEO). As of November 2013, one company, SpaceX, is two years along on a privately funded multi-year development program for a reusable rocket launching system with the stated intention to commercialize "fully and rapidly reusable" launch technology. SpaceX has completed eight test flights of their low-altitude booster return prototype, Grasshopper, and one test flight of a high-altitude/high-velocity booster return test vehicle, with a second booster return test flight planned for early 2014.

Much of the material launched need not be delivered to its eventual orbit immediately, which raises the possibility that high efficiency (but slower) engines could move SPS material from LEO to GEO at an acceptable cost. Examples include ion thrusters or nuclear propulsion.

Power beaming from geostationary orbit by microwaves carries the difficulty that the required 'optical aperture' sizes are very large. For example, the 1978 NASA SPS study required a 1-km diameter transmitting antenna, and a 10 km diameter receiving rectenna, for a microwave beam at 2.45 GHz. These sizes can be somewhat decreased by using shorter wavelengths, although they have increased atmospheric absorption and even potential beam blockage by rain or water droplets. Because of the thinned array curse, it is not possible to make a narrower beam by combining the beams of several smaller satellites. The large size of the transmitting and receiving antennas means that the minimum practical power level for an SPS will necessarily be high; small SPS systems will be possible, but uneconomic.

To give an idea of the scale of the problem, assuming a solar panel mass of 20 kg per kilowatt (without considering the mass of the supporting structure, antenna, or any significant mass reduction of any focusing mirrors) a 4 GW power station would weigh about 80,000 metric tons, all of which would, in current circumstances, be launched from the Earth. Very lightweight designs could likely achieve 1 kg/kW, meaning 4,000 metric tons for the solar panels for the same 4 GW capacity station. This would be the equivalent of between 40 and 150 heavy-lift launch vehicle (HLLV) launches to send the material to low earth orbit, where it would likely be converted into subassembly solar arrays, which then could use high-efficiency ion-engine style rockets to (slowly) reach GEO (Geostationary orbit). With an estimated serial launch cost for shuttle-based HLLVs of $500 million to $800 million, and launch costs for alternative HLLVs at $78 million, total launch costs would range between $11 billion (low cost HLLV, low weight panels) and $320 billion ('expensive' HLLV, heavier panels). To these costs must be added the environmental impact of heavy space launch emissions, if such costs are to be used in comparison to earth-based energy production. For comparison, the direct cost of a new coal or nuclear power plant ranges from $3 billion to $6 billion per GW (not including the full cost to the environment from CO2 emissions or storage of spent nuclear fuel, respectively); another example is the Apollo missions to the Moon cost a grand total of $24 billion (1970s' dollars), taking inflation into account, would cost $140 billion today, more expensive than the construction of the International Space Station.

However in 2013 based on Recent innovations, Electric Space: Space-Based Solar Power Technologies & Applications suggested a new way to reduce costs by replacing smaller satellites and in lower Orbits.

Non-Conventional Launch Methods

SBSP costs could be significantly reduced if a means of putting the necessary materials into orbit were developed that did not rely on expendable rockets. One idea for this is the non-rotating Skyhook, a type of space elevator that could be built with existing materials and technology. Some more speculative possibilities include ground launch systems such as Star Tram, mass drivers or launch loops, or the Earth surface to geostationary orbit Space Elevator. However, these ideas require technologies that have yet to be developed.

Building from Space

From Lunar Materials Launched in Orbit

Gerard O'Neill, noting the problem of high launch costs in the early 1970s, proposed building the SPS's in orbit with materials from the Moon. Launch costs from the Moon are potentially much lower than from Earth, due to the lower gravity. This 1970s proposal assumed the then-advertised future launch costing of NASA's space shuttle. This approach would require substantial up front capital investment to establish mass drivers on the Moon.

Nevertheless, on 30 April 1979, the Final Report ("Lunar Resources Utilisation for Space Construction") by General Dynamics' Convair Division, under NASA contract NAS9-15560, concluded that use of lunar resources would be cheaper than Earth-based materials for a system of as few as thirty Solar Power Satellites of 10GW capacity each.

In 1980, when it became obvious NASA's launch cost estimates for the space shuttle were grossly optimistic, O'Neill et al. published another route to manufacturing using lunar materials with much lower startup costs. This 1980s SPS concept relied less on human presence in space and more on partially self-replicating systems on the lunar surface under remote control of workers stationed on Earth. The high net energy gain of this proposal derives from the Moon's much shallower gravitational well. Having a relatively cheap per pound source of raw materials from space would lessen the concern for low mass designs and result in a different sort of SPS being built. The low cost per pound of lunar materials in O'Neill's vision would be supported by using lunar material to manufacture more facilities in orbit than just solar power satellites. Advanced techniques for launching from the Moon may reduce the cost of building a solar power satellite from lunar materials. Some proposed techniques include the lunar mass driver and the lunar space elevator, first described by Jerome Pearson. It would require establishing silicon mining and solar cell manufacturing facilities on the Moon.

On the Moon

David Criswell suggests the Moon is the optimum location for solar power stations, and promotes lunar solar power. The main advantage he envisions is construction largely from locally available lunar materials, using in-situ resource utilisation, with a teleoperated mobile factory and crane to assemble the microwave reflectors, and rovers to assemble and pave solar cells, which would significantly reduce launch costs compared to SBSP designs. Power relay satellites orbiting around earth and the Moon reflecting the microwave beam are also part of the project. A demo project of 1 GW starts at $50 billion. The Shimizu Corporation use combination of lasers and microwave for the lunar ring concept, along with power relay satellites.

From an Asteroid

Asteroid mining has also been seriously considered. A NASA design study evaluated a 10,000 ton mining vehicle (to be assembled in orbit) that would return a 500,000 ton asteroid fragment to geostationary orbit. Only about 3,000 tons of the mining ship would be traditional aerospace-grade payload. The rest would be reaction mass for the mass-driver engine, which could be arranged to be the spent rocket stages used to launch the payload. Assuming that 100% of the returned asteroid was useful, and that the asteroid miner itself couldn't be reused, that represents nearly a 95% reduction in launch costs. However, the true merits of such a method would depend on a thorough mineral survey of the candidate asteroids; thus far, we have only estimates of their composition. One proposal is to capture the asteroid Apophis into earth orbit and convert it into 150 solar power satellites of 5 GW each.

Solar Panels on Spacecraft

Spacecraft operating in the inner solar system usually rely on the use of photovoltaic solar panels to derive electricity from sunlight. In the outer solar system, where the sunlight is too weak to produce sufficient power, radioisotope thermal generators (RTGs) are used as a power source.

The first spacecraft to use solar panels was the Vanguard 1 satellite, launched by the US in 1958. This was largely because of the influence of Dr. Hans Ziegler, who can be regarded as the father of spacecraft solar power. Of his 30 year tenure at Fort Monmouth (1947–1976), Ziegler spent 12 years in the top position as Chief Scientist.

Uses

Solar panels on spacecraft supply power for 2 main uses:

- power to run the sensors, active heating, cooling and telemetry.
- power for spacecraft propulsion — electric propulsion, sometimes called solar-electric propulsion.

For both uses, the figure of merit of the solar panels (and RTGs) is the power generated per kilogram, as an upper limit of the power the spacecraft has at its disposal per kg spacecraft (including solar panels). To increase the power generated per kg, typical solar panels on spacecraft use close-packed solar cell rectangles that cover nearly 100% of the sun-visible area of the solar panels, rather than the solar wafer circles which, even though close-packed, cover about 90% of the sun-visible area of typical solar panels on earth. However, some solar panels on spacecraft have solar cells that cover only 30% of the sun-visible area.

Implementation

Solar panels need to have a lot of surface area that can be pointed towards the Sun as the spacecraft moves. More exposed surface area means more electricity can be converted from light energy from the Sun. Since spacecraft have to be small, this limits the amount of power that can be produced.

All electrical circuits generate waste heat; in addition, solar arrays act as optical and thermal as well as electrical collectors. Heat must be radiated from their surfaces. High-power spacecraft may have solar arrays that compete with the active payload itself for thermal dissipation. The innermost panel of arrays may be "blank" to reduce the overlap of views to space. Such spacecraft include the higher-power communications satellites (e.g., later-generation TDRS) and Venus Express, not high-powered but closer to the Sun.

Spacecraft are built so that the solar panels can be pivoted as the spacecraft moves. Thus, they can always stay in the direct path of the light rays no matter how the spacecraft is pointed. Spacecraft are usually designed with solar panels that can always be pointed at the Sun, even as the rest of the body of the spacecraft moves around, much as a tank turret can be aimed independently of where the tank is going. A tracking mechanism is often incorporated into the solar arrays to keep the array pointed towards the sun. Sometimes, satellite operators purposefully orient the solar panels to "off point," or out of direct alignment from the Sun. This happens if the batteries are completely charged and the amount of electricity needed is lower than the amount of electricity made; off-pointing is also sometimes used on the International Space Station for orbital drag reduction.

Types of Solar Cells Typically Used

Gallium arsenide-based solar cells are typically favoured over silicon in industry, because they have a higher efficiency. The most efficient solar cells currently in production are multi-junction cells. These use a combination of several layers of both gallium arsenide and silicon to capture the largest spectrum of light possible. Leading edge multi-junction cells are capable of nearly 29% efficiency under ideal conditions.

Spacecraft that have Used Solar Power

To date, solar power, other than for propulsion, has been practical for spacecraft operating no farther from the Sun than the orbit of Jupiter. For example Juno, Magellan, Mars Global Surveyor, and Mars Observer used solar power as does the Earth-orbiting, Hubble Space Telescope. The Rosetta space probe, launched March 2, 2004, will use solar panels as far as the orbit of Jupiter (5.25 AU); previously the furthest use was the Stardust spacecraft at 2 AU. Solar power for propulsion was also used on the European lunar mission SMART-1 with a Hall effect thruster.

The *Juno* mission is the first mission to Jupiter to use solar panels instead of the traditional RTGs that are used by previous outer solar system missions. Another spacecraft of interest is Dawn which went into orbit around 4 Vesta in 2011. The potential for solar powered spacecraft beyond Jupiter has been studied.

Power Available

In 2005 Rigid-Panel Stretched Lens Arrays were producing 7 kW per wing. Solar arrays producing 300 W/kg and 300 W/m^2 from the sun's 1366 W/m^2 power near the Earth are available. Entech Inc. hopes to develop 100 kW panels by 2010 and 1 MW panels by 2015.

Future Uses

For future missions, it is desirable to reduce solar array mass, and to increase the power generated per unit area. This will reduce overall spacecraft mass, and may make the operation of solar-powered spacecraft feasible at larger distances from the sun. Solar array mass could be reduced with thin-film photovoltaic cells, flexible blanket substrates, and composite support structures. Solar array efficiency could be improved by using new photovoltaic cell materials and solar concentrators that intensify the incident sunlight. Photovoltaic concentrator solar arrays for primary spacecraft power are devices which intensify the sunlight on the photovoltaics. This design uses a flat lens, called a Fresnel lens, which takes a large area of sunlight and concentrates it onto a smaller spot. The same principle is used to start fires with a magnifying glass on a sunny day.

Solar concentrators put one of these lenses over every solar cell. This focuses light from the large concentrator area down to the smaller cell area. This allows the quantity of expensive solar cells to be reduced by the amount of concentration. Concentrators work best when there is a single source of light and the concentrator can be pointed right at it. This is ideal in space, where the Sun is a single light source. Solar cells are the most expensive part of solar arrays, and arrays are often a very expensive part of the spacecraft. This technology may allow costs to be cut significantly due to the utilisation of less material.

It has been proposed that it may be possible to develop space-based solar plants — solar power satellites with large arrays of photovoltaic cells—that would beam the energy they produce to Earth using microwaves or lasers. This could, in principle, be a significant source of electrical power generated using non-fossil fuel sources. Japanese and European space agencies, among others, are analyzing the possibility of developing such power plants in the 21st century.

Space and Concentrator Systems

Space solar power satellites require innovative concepts in order to achieve economically and technically feasible designs. The mass and volume constraints of current and planned launch vehicles necessitate highly efficient structural systems be developed. In addition, modularity and in-space deployment will be enabling design attributes. For four decades, the concept of deriving terrestrial energy from space-based solar-electric systems using wireless power transfer has captured the imagination of government and private stakeholders. Various studies of this concept were conducted during the 1970s, by NASA and the Department of Energy.

In 1995 NASA's Advanced Concepts Office initiated a "fresh look" at the requirements and technology for a space solar power (SSP) system.[3] The Mankins' study identified new system concepts including the "SunTower" - a gravity gradient stabilized, space tether-based SSP system concept. In the late 1990's, NASA developed a new SPS concept based on optical concentrators. The Integrated Symmetrical Concentrator concept utilises thin film optics to concentrate the solar radiation and thereby reduces the photo-voltaic array size. More recently the National Security Space Office (NSSO) studied the use of Space-Based Solar Power. The NSSO study also adopted the symmetrical concentrator SPS concept. In each of the SSP concepts studied to date, very large structural systems are an enabling attribute. Since the system cost is highly dependent on mass and complexity, development of advanced mechanical systems is crucial to achieving economical designs.

In this paper, a two-phase building block approach is proposed to develop and validate large SSP systems. The first building block is a near-term, low power, tactical system with application to customers willing to pay a premium for consistent and uninterrupted power. This smaller low power system would validate fundamental technologies as well as technical and operational models. The proposed second phase of SSP development would involve focused investments in advanced technologies required for systems capable of producing commercial levels of power and transferring that to the terrestrial power grid. The materials, structures, and mechanical systems (MSMS) technologies that can be employed in a near term tactical demonstrator and for very large SPS systems are briefly described.

Building Block Approach: A Two Phase SSP Development

The first phase focuses on a near term application/customer of SSP that is willing to pay a premium for consistent and uninterrupted power. An example of this would be military bases in remote and hostile regions, where the logistics train for fuel (to run generators) is very expensive, dangerous, and subject to constant disruption. Low power SSP systems may also be used in orbit around the Moon, Mars and other solar system planets and moons to provide power to surface rovers and outposts. The power generation level (at the source) for this first phase application might be from 100-5000 KW. This application would use current and near term technology (structures, solar cells, ion propulsion/station keeping, avionics and power beaming) for spacecraft subsystems and automated rendezvous and docking for spacecraft assembly.

The goal of the second phase would be to develop the advanced technologies required for SSP systems capable of producing 100-2000 MW of power for commercial transfer to the Earth's power grid. Such large satellites would only be developed when appropriate systems and technologies were sufficiently advanced to make them commercially viable. Using block upgrades on first phase systems to develop and demonstrate the advanced technologies as they become available would reduce the cost, schedule and performance risks of very large system implementation. In addition, the probability of commercial system development success would be maximized because development would not begin prematurely.

General design features that significantly contributed to the SETV configuration applicable to a first phase SPS include:

1. the system will be composed of modular units that can be replaced in situ;

2. the configuration will be amenable to a variety of design implementations and system decomposition, allowing it to be packaged on a variety of launch vehicles; and,
3. the launched components will be capable of being assembled on orbit using both human and robotic resources and capabilities.

Individual sub-modules and components that make up one 50 KW-class SETV System Module are show in Figure. The reference 450 KW-class SETV is composed of eight of these SETV System Modules. With a few modifications, the near term SETV could form the basis of the first phase SPS system. Instead of all of the power in each SETV module being routed to its associated thrusters, the bulk would be routed to a wireless power transmitter (microwave or laser) mounted at the centre of the spacecraft for transmission to the Earth's surface. The system was sized such that a 50 KW-class module could be launched by a Delta 2-Heavy, or four pre-assembled modules launched in a Delta 4-Heavy class launch vehicle. Automated on-orbit rendezvous and docking is an established technology and is used to dock modules to assemble the complete SPS system. The system could be assembled in LEO, to allow for checkout, and then it could propel itself to its service orbit.

Second Phase SPS System

The successful first phase SPS demonstrator will provide the knowledge in terms of validated performance/economics models and operations experience to permit large scale system architectures to be developed for a 1000 MW class SPS. The Integrated Symmetrical Concentrator Concept is one of those promising architectures. The choice of wireless power transfer technology, specifically the wavelength (RF or Laser), can have a substantial influence on the SPS antenna size and thermal requirements. Moreover, large inflatable concentrators have been proposed to reduce the photo-voltaic area (and cost), but little attention has been paid to long term space durability. These geometric and environmental requirements necessitate the continued advancement of materials, structures and mechanical systems (MSMS) for the second phase: 1000 MW class SPS. To this end, specific advances in MSMS technology that would benefit very large space structural systems in the coming decades are discussed in the following section.

Materials, Structures and Mechanical Systems (MSMS) Technology

The MSMS technology areas selected for this discussion include:

- Modularity (Module Based Assembly and Upgrade)
- Material Systems (Space Durable, High Temperature, and Thin Films)
- Structural Concepts (Inflatable, Rigidizable and Gossamer Concepts)
- In-Space Operations (Deployment, Assembly and Repair)

Modularity

The goal of modularity is to simplify space-platform design by developing versatile repeating units that have a range of common features and interfaces. Modularity reduces mission risk, and allows spares and replacements to be available during system assembly. Modularity, together with a robust capability to perform in-space assembly, has the potential to greatly enable the SSP mission. The modular assembly approach was used as a basis for designing the previously described Solar Electric Transfer Vehicle (SETV).

The modular assembly design approach is significantly more complex than to simply design repeating and versatile units. The complete and comprehensive modular assembly system-of-systems encompass all of the following:

1. the mission-level system that utilises the modules and the associated module specifications and designs;
2. the agents that assemble, service and repair the modular subsystems (robots or astronauts);
3. the operations, and associated planning, required during assembly (positioning, aligning, joining), servicing and repair; and
4. the infrastructure (cranes and mobile platforms) required to facilitate operations and enhance agent capabilities.

All of these systems must be considered simultaneously in order to develop a modular architecture that maximizes the potential benefits.

Material Systems

Materials are enabling for any structural system. For SPS systems, lightweight materials are needed that exhibit the proper combination of mechanical, thermal, and electrical properties for long periods of time in the space environment, including radiation and micrometeroid and orbital debris (MMOD) exposure. Space durable and high temperature material systems with high specific strength and stiffness are needed both for the primary structure and for devices and sensors. A survey by Harris, et al.[7] in 2002 identified aerospace materials in current use and those with high potential for the future. The study indicated that single wall carbon nanotube based materials offer the potential for a three to five fold increase in specific stiffness and nearly two orders of magnitude increase in specific strength as compared to state-of-the-art polymer matrix composites. Much research has occurred to advance the knowledge of carbon nanotube materials (CNT) and more recently Boron nitride nanotubes (BNT) materials. Both CNT and BNT materials are attractive for application to SPS because of their unique thermal, electrical, and mechanical properties.

In addition to high stiffness and strength properties, materials conditioned to undergo large strains during deployment and capable of post-deployment rigidization are highly desired to facilitate packaging for launch and deployment of SPS systems on-orbit. Finally, damage due to MMOD impact on large systems such as SPS concentrators and arrays makes the development of materials with self-healing properties critical to achieve long term operations. In particular, recent and future advances in high temperature materials, rigidization technology and self-healing materials will provide lower cost, longer life SSP systems.

Structural Concepts

A high level metric during the "Fresh Look Study" was to achieve a structural design for the primary SPS structure at less than < 3-4 kg/KW. For a 1000 MW SPS, considering the launch cost to be $1000/kg, this would require an investment of $3-4 billion to launch just the primary structure. For most aerospace systems, the primary structure is approximately 20% of the dry mass. Thus, a 1000 MW SPS would require $15-20 billion for launch costs alone. Clearly, structural concepts that enable a lightweight architecture will have a first-order impact on SPS economics. In addition to low mass, the structural concepts must also support a flexible and modular approach to achieving SPS long-life (15+ years).

The integrated symmetrical concentrator SPS concept requires multi-kilometre size concentrator/reflector arrays. While these structural components will be gossamer thin film tension stabilized membranes, they will also require shape control to keep solar light reasonably focused (in a non-coherent manner) on the photovoltaic arrays. Thus, the concentrator designs will be both mass and stiffness driven designs utilising high performance materials. NASA solar sail technology investments over the past decade have provided lightweight structural concepts directly applicable to SPS concentrators.

In-Space Operations

In order to field a SSP system, a variety of in-space operational capabilities are needed. Since each SPS will be very large, it is assumed that multiple launches will be required to place the subsystems into low Earth orbit (LEO). Two options exist for completing the system, either assembling in LEO and then transferring the completed system to its final orbit, or transporting all of the subsystems to the final orbit and performing final assembly there. Either approach will require a robust set of in-space operational capabilities, including automated rendezvous, docking and berthing, assembly, and servicing and repair. Recent robotics missions have significantly matured the key in-space operations technologies needed for SSP.

Solar Cell Efficiency

Solar cell efficiency is the ratio of the electrical output of a solar cell to the incident energy in the form of sunlight. The energy conversion efficiency (η) of a solar cell is the percentage of the solar energy to which the cell is exposed that is converted into electrical energy.

This is calculated by dividing a cell's power output (in watts) at its maximum power point (P_m) by the input light (E, in W/m^2) and the surface area of the solar cell (A_c in m^2).

$$\eta = \frac{P_m}{E \times A_c}$$

By convention, solar cell efficiencies are measured under standard test conditions (STC) unless stated otherwise. STC specifies a temperature of 25 °C and an irradiance of 1000 W/m^2 with an air mass 1.5 (AM1.5) spectrum.

These conditions correspond to a clear day with sunlight incident upon a sun-facing 37°-tilted surface with the sun at an angle of 41.81° above the horizon. This represents solar noon near the spring and autumn equinoxes in the continental United States with surface of the cell aimed directly at the sun. Under these test conditions a solar cell of 20% efficiency with a 100 cm^2 (0.01 m^2) surface area would produce 2.0 W.

The efficiency of the solar cells used in a photovoltaic system, in combination with latitude and climate, determines the annual energy output of the system. For example, a solar panel with 20% efficiency and an area of 1 m^2 will produce 200 W at STC, but it can produce more when the sun is high in the sky and will produce less in cloudy conditions and when the sun is low in the sky.

In central Colorado, which receives daily insolation of 2200 Wh/m^2, such a panel can be expected to produce 440 kWh of energy per year. However, in Michigan, which receives only 1400 kWh/m^2/yr, annual energy yield will drop to 280 kWh for the same panel. At more northerly European latitudes, yields are significantly lower: 175 kWh annual energy yield in southern England.

Several factors affect a cell's conversion efficiency value, including its reflectance efficiency, thermodynamic efficiency, charge carrier separation efficiency, and conduction efficiency values.

Because these parameters can be difficult to measure directly, other parameters are measured instead, including quantum efficiency, V_{OC} ratio, and fill factor. Reflectance losses are accounted for by the quantum efficiency value, as they affect "external quantum efficiency." Recombination losses are accounted for by the quantum efficiency, V_{OC} ratio, and fill factor values.

Resistive losses are predominantly accounted for by the fill factor value, but also contribute to the quantum efficiency and V_{OC} ratio values. In 2014, the highest efficiencies have been achieved by using multiple junction cells at high solar concentrations (44.7% by the Fraunhofer ISE, Soitec, CEA-Leti and the Helmholtz Centre Berlin

Factors Affecting Energy Conversion Efficiency

The maximum theoretically possible conversion efficiency for sunlight is 86% due to the entropy of the photons emitted by the sun's surface.

However, solar cells operate as quantum energy conversion devices, and are therefore subject to the "thermodynamic efficiency limit". Photons with an energy below the band gap of the absorber material cannot generate a hole-electron pair, and so their energy is not converted to useful output and only generates heat if absorbed.

For photons with an energy above the band gap energy, only a fraction of the energy above the band gap can be converted to useful output. When a photon of greater energy is absorbed, the excess energy above the band gap is converted to kinetic energy of the carrier combination. The excess kinetic energy is converted to heat through phonon interactions as the kinetic energy of the carriers slows to equilibrium velocity.

Solar cells with multiple band gap absorber materials improve efficiency by dividing the solar spectrum into smaller bins where the thermodynamic efficiency limit is higher for each bin.

Quantum Efficiency

The term quantum efficiency (QE) may apply to incident photon to converted electron (IPCE) ratio, of a photosensitive device or it may refer to the TMR effect of a Magnetic Tunnel Junction. In a charge-coupled device (CCD) is the percentage of photons hitting the device's photoreactive surface that produce charge carriers. It is measured in electrons per photon or amps per watt. QE is a measurement of a device's electrical sensitivity to light. Since the energy of a photon is inversely proportional to its wavelength, QE is often measured over a range of different wavelengths to characterize a device's efficiency at each photon energy level. The QE for photons with energy below the band gap is zero. Photographic film typically has a QE of much less than 10%, while CCDs can have a QE of well over 90% at some wavelengths.

A solar cell's quantum efficiency value indicates the amount of current that the cell will produce when irradiated by photons of a particular wavelength. If the cell's quantum efficiency is integrated over the whole solar electromagnetic spectrum, one can evaluate the amount of current that the cell will produce when exposed to sunlight. The ratio between this energy-production value and the highest possible energy-production value for the cell (i.e., if the QE were 100% over the whole spectrum) gives the cell's overall energy conversion efficiency value. Note that in the event of multiple exciton generation (MEG), quantum efficiencies of greater than 100% may be achieved since the incident photons have more than twice the band gap energy and can create two or more electron-hole pairs per incident photon.

Types of Quantum Efficiency

Two types of quantum efficiency of a solar cell are often considered:

- External Quantum Efficiency (EQE) is the ratio of the number of charge carriers collected by the solar cell to the number of photons of a given energy *shining on the solar cell from outside* (incident photons).
- Internal Quantum Efficiency (IQE) is the ratio of the number of charge carriers collected by the solar cell to the number of photons of a given energy that shine on the solar cell from outside *and* are absorbed by the cell.

The IQE is always larger than the EQE. A low IQE indicates that the active layer of the solar cell is unable to make good use of the photons. To measure the IQE, one first measures the EQE of the solar device, then measures its transmission and reflection, and combines these data to infer the IQE.

$$\text{EQE} = \frac{\text{electrons/sec}}{\text{photons/sec}} = \frac{\text{current / (charge of 1 electron)}}{\text{(total power of photons) / (energy of one photon)}}$$

$$\text{IQE} = \frac{\text{electrons/sec}}{\text{absorbed photons/sec}} = \frac{\text{EQE}}{\text{1-Reflection-Transmission}}$$

The external quantum efficiency therefore depends on both the absorption of light and the collection of charges. Once a photon has been absorbed and has generated an electron-hole pair, these charges must be separated and collected at the junction. A "good" material avoids charge recombination. Charge recombination causes a drop in the external quantum efficiency.

The ideal quantum efficiency graph has a square shape, where the QE value is fairly constant across the entire spectrum of wavelengths measured. However, the QE for most solar cells is reduced because of the effects of recombination, where charge carriers are not able to move into an external circuit. The same mechanisms that affect the collection probability also affect the QE. For example, modifying the front surface can affect carriers generated near the surface. And because high-energy (blue) light is absorbed very close to the surface, considerable recombination at the front surface will affect the "blue" portion of the QE. Similarly, lower energy (green) light is absorbed in the bulk of a solar cell, and a low diffusion length will affect the collection probability from the solar cell bulk, reducing the QE in the green portion of the spectrum. Generally, solar cells on the market today do not produce much electricity from ultraviolet and infrared light (<400 nm and >1100 nm wavelengths, respectively); this light is either filtered out or absorbed by the cell, heating the cell. That heat is wasted energy, and could lead to damage to the cell.

Quantum efficiency of Image Sensors : Quantum efficiency (QE) is the fraction of photon flux that contributes to the photocurrent in a photodetector or a pixel. Quantum efficiency is one of the most important parameters used to evaluate the quality of a detector and is often called the spectral response to reflect it's wavelength dependence. It is defined as the number of signal electrons created per incident photon. In some cases it can exceed 100% (i.e. more than one electron created per incident photon).

Spectral Responsivity

Spectral responsivity is a similar measurement, but it has different units: amperes per watt (A/W); (i.e. how much current comes out of the device per incoming photon of a given energy and wavelength). Both the quantum efficiency and the responsivity are functions of the photons' wavelength (indicated by the subscript λ).

To convert from responsivity (R_λ, in A/W) to QE_λ (on a scale 0 to 1):

$$QE_\lambda = \frac{R_\lambda}{\lambda} \times \frac{hc}{e} \approx \frac{R_\lambda}{\lambda} \times (1240\ \text{W}\cdot\text{nm/A})$$

where λ is in nm, h is the Planck constant, c is the speed of light in a vacuum, and e is the elementary charge.

Determination

$$QE_\lambda = \eta = \frac{N_e}{N_\nu}$$

where N_e = number of electrons produced, N_ν = number of photons absorbed.

$$\frac{N_\nu}{t} = \Phi_o \frac{\lambda}{hc}$$

Assuming each photon absorbed in the depletion layer produces a viable electron-hole pair, and all other photons do not,

$$\frac{N_e}{t} = \Phi_\xi \frac{\lambda}{hc}$$

where t is the measurement time (in seconds), Φ_o = incident optical power in watts, Φ_ξ = optical power absorbed in depletion layer, also in watts.

Maximum Power Point

A solar cell may operate over a wide range of voltages (V) and currents (I). By increasing the resistive load on an irradiated cell continuously from zero (a *short circuit*) to a very high value (an *open circuit*) one can determine the maximum power point, the point that maximizes V×I; that is, the load for which the cell can deliver maximum electrical power at that level of irradiation. (The output power is zero in both the short circuit and open circuit extremes).

A high quality, monocrystalline silicon solar cell, at 25 °C cell temperature, may produce 0.60 V open-circuit (V_{OC}). The cell temperature in full sunlight, even with 25 °C air temperature, will probably be close to 45 °C, reducing the open-circuit voltage to 0.55 V per cell. The voltage drops modestly, with this type of cell, until the short-circuit current is approached (I_{SC}). Maximum power (with 45 °C cell temperature) is typically produced with 75% to 80% of the open-circuit voltage (0.43 V in this case) and 90% of the short-circuit current. This output can be up to 70% of the V_{OC} x I_{SC} product. The short-circuit current (I_{SC}) from a cell is nearly proportional to the illumination, while the open-circuit voltage (V_{OC}) may drop only 10% with an 80% drop in illumination. Lower-quality cells have a more rapid drop in voltage with increasing current and could produce only 1/2 V_{OC} at 1/2 I_{SC}. The usable power output could thus drop from 70% of the V_{OC} x I_{SC} product to 50% or even as little as 25%. Vendors who rate their solar cell "power" only as V_{OC} x I_{SC}, without giving load curves, can be seriously distorting their actual performance.

The maximum power point of a photovoltaic varies with incident illumination. For example, accumulation of dust on photovoltaic panels reduces the maximum power point. For systems large enough to justify the extra expense, a maximum power point tracker tracks the instantaneous power by continually measuring the voltage and current (and hence, power transfer), and uses this information to dynamically adjust the load so the maximum power is *always* transferred, regardless of the variation in lighting.

Bibliography

Altmann, J. & Gubrud, M.A.: *Discovering the Nanoscale*, IOS Press, Amsterdam, 2004.

Andrew, N.: *Foundations of Nanomechanics: from Solid-state Theory to Device Applications*, Berlin, New York, 2003.

Attebery, B.: *Nanoculture: Implications of the New Technoscience*, Intellect Books, Bristol, 2004.

Bainbridge, W.S. : *Societal Implications of Nanoscience and Nanotechnology*, Dordrecht, Kluwer, 2001.

Bast, R.C. Jr.: *Cancer Medicine*, Decker, Hamilton, 2000.

Berube, D.M.: *Discovering the Nanoscale*, IOS Press, Amsterdam, 2004.

Brosnan, J.: *The Primal Scream: A History of Science Fiction Film*, Orbit Books, London, 1991.

Carroll, J.S.: *Social Science Research Methods for Assessing Societal Implications of Nanotechnology*, Dordrecht, Kluwer, 2001.

Caviedes, C.: *El Nino in History: Storming Through the Ages*, University Press of Florida, Florida, 2001.

Credi, A.: *Molecular Devices and Machines: a Journey into the Nano World*, Wiley-VCH, Weinheim, 2003.

Crow, M. & Sarewitz, D.: *Societal Implications of Nanoscience and Nanotechnology*, Dordrecht: Kluwer, 2001.

Daniel Ratner: *Nanotechnology: a Gentle Introduction to the Next Big Idea*, Upper Saddle River, Prentice Hall, 2003.

Dobzhanshy, T.: *Genetics of the Evolutionary Process*, Columbia University Press, New York, 1970.

Drexler, K. Eric: *Engines of Creation: The Coming Era of Nanotechnology*, Anchor Books, New York, 1986.

Dyson, Tim: *India's Historical Demography: Studies in Famine, Disease, and Society*, Riverdale, Westwood, MA, 1989.

Eamonn Gearon: *The Sahara: A Cultural History*, Signal Books, UK, 2011.

Frank J. Owens: *Introduction to Nanotechnology*, Hoboken, J. Wiley, 2003.

Glimell, H.: *Dynamics of the Emerging Field of Nanoscience*, Dordrecht, Kluwer, 2001.

Glimell, H.: *Societal Implications of Nanoscience and Nanotechnology*, Kluwer, Dordrecht, 2001.

Haber, L. F.: *The Chemical Industry: 1900-1930, International Growth and Technological Change*, Oxford: Claredon Press, 1971.

Hessenbruch, A.: *Nanotechnology and the Negotiation of Novelty*, Amsterdam, IOS Press, 2004.

James Lewis: *Prospects in Nanotechnology: Toward Molecular Manufacturing,* John Wiley & Sons, New York, 1995.

Karn, Barbara: *Nanotechnology and the Environment: Applications and Implications,* American Chemical Society, Washington D.C., 2004.

Keblinski, P. : *Encyclopedia of Nanoscience and Nanotechnology*, American Scientific, Los Angeles, Calif, USA, 2004.

Liu W.K., Karpov E.G., Park H.S.: *Nano Mechanics and Materials,* Wiley, Weinheim, 2006.

Mayr, E.: *Change of Genetic Environment and Evolution*, Allen and Unwin, London, 1954.

Mody, C.C.M.: *How Probe Microscopists Became Nanotechnologists, Discovering the Nanoscale*, IOS Press, Amsterdam, 2004.

Norio Taniguchi: *Nanotechnology: Integrated Processing Systems for Ultra-precision and Ultra-fine Products,* Oxford University Press, New York, 1996.

Orth, R. N.: *Nanometer Scale Antibody Patterning for Directed cell Immobilization and Stimulation,* Cornell University, Ithaca, 2002.

Pressman, J.: *Nano Narrative: A Parable from Electronic Literature*, Intellect Books, Bristol, 2004.

Pressman, J.: *Nanoculture: Implications of the New Technoscience*, Intellect Books, Bristol, UK, 2004.

Ratner, Daniel: *Nanotechnology: a Gentle Introduction to the Next Big Idea*, Prentice Hall, Upper Saddle River, 2003.

Robison, W.L.: *Discovering the Nanoscale*, IOS Press, Amsterdam, 2004.

Sarewitz, D.: *Nanotechnology and Societal Transformation*, Dordrecht, Kluwer, 2001.

Stephen M. Goodnick: *Transport in Nanostructures,* Cambridge University Press, Cambridge, 1997.

Tenner, E.: *Nanotechnology and Unintended Consequences*, , Dordrecht, Kluwer, 2001.

Vasudev Bhatnagar: *Cell Science and Technology*, Campus Books, Delhi, 2009.

Vermaas, P.E.: *Discovering the Nanoscale*, IOS Press, Amsterdam, 2004.

Weil, V.: *Ethical Issues in Nanotechnology*, Dordrecht, Kluwer, 2001.

Index

L

M

N

O

P

Q

R

S

T

V

W

❑❑❑